Algebra Answer Book

Solutions to every question

—

Step by step :)

I'm stuck again.

I don't even know what I'm doing wrong.

My answer's not even close to what's at the back.

It's doing my head in.

Yeah, that's annoying. But we all get stuck.

It's part of the learning process.

So, what am I supposed to do?

Continue to feel dumb and hopeless?

No. Of course not.

Next time you get stuck, do this.

Reread the explanations, and go through the worked examples.

After that, on the question you're stuck on, spend about 5-10 minutes on it.

Tops.

If you still don't get it, look at the Answer Book.

Yeah, I got that with the original book.

Cool. That's why it's there.

Find the question you got stuck on.

Check how it was answered.

Understand it.

Then go back and try the question again.

If you get it right, great.

Move on to the next question.

If not, look at the answer again.

Pinpoint your mistake.

Where are you going wrong?

Learn from it.

Then try the question again, until you get it right.

It's the fastest way you'll learn.

Why didn't anyone tell me this before?

Beats me.

But you know now.

The only thing is, if you're checking the answer book a bit too much.

There's something about the topic you don't get.

Maybe from the topic you're on.

Maybe something from earlier.

Either way, go back to where things last made sense,

and rebuild back up from there.

Find what you're not getting.

Fix it.

This way you're always making sure you really understand things,

before moving on.

And hopefully you won't be stuck as much.

Stuff You're About to Learn How to Answer

Linear Equations

Simultaneous Equations

The Basics

Divisibility Rules

Multiple Choice

1 **Option A**

$7 + 1 + 4 = 9$

9 is divisible by 3.

∴ 714 is divisible by 3.

2 **Option C**

Divisible by 10: Last digit must be 0.

3 **Option D**

528 is even.

$5 + 2 + 8 = 15$

15 is divisible by 3.

∴ 528 is divisible by 6.

4 **Option B**

$4 + 6 + 8 = 18$

18 is divisible by 9.

∴ 468 is divisible by 9.

Missing digit is 6

5 **Option D**

The last two digits are 36.

36 is divisible by 4.

∴ Any first digit will make number divisible by 4.

Time To Practice

1 24

Last digit is even

$\rightarrow 4$ divisible by 2

$\therefore$ Yes

2 42

Sum of digits $4 + 2 = 6$

$\rightarrow 6$ divisible by 3

$\therefore$ Yes

3 45

Last digit is 5

$\rightarrow 5$ divisible by 5

$\therefore$ Yes

4 68

Last digit is even.

Sum of digits $6 + 8 = 14$

$\rightarrow$ 14 NOT divisible by 3

$\therefore$ No

5 72

Using normal division $72 \div 8 = 9$

$\rightarrow$ 72 divisible by 8

$\therefore$ Yes

6 84

Last digit is even

Sum of digits $8 + 4 = 12$

$\rightarrow$ 12 is divisible by 3

$\therefore$ Yes

7 84

$8 - (2 \times 4) = 0$

$\rightarrow 0$

$\therefore$ Yes

8 94

$9 - (2 \times 4) = 1$

$\rightarrow$ 1 NOT divisible by 7

$\therefore$ No

9 98

Last two digit 98

$\rightarrow$ 98 is NOT divisible by 4

$\therefore$ No

10 105

Sum of digits $1 + 0 + 5 = 6$

$\rightarrow$ 6 is NOT divisible by 9

$\therefore$ No

11 114

$11 - (2 \times 4) = 3$

→ 3 NOT divisible by 7

∴ No

12 144

Last two digit 44

→ 44 is divisible by 4

∴ Yes

13 120

a **2:** Last digit is even → divisible by 2

3: Sum of digits $1 + 2 + 3 = 68$ → divisible by 3

4: Last 2 is digits 20 $(20 \div 4 = 5)$ → is divisible by 4

5: Last digit is 0 → divisible by 5

6: Even and divisible by 3 → divisible by 6

7: $12 - (2 \times 0) = 12$ → NOT divisible by 7

8: Last three digits $120 \div 8 = 15$ → divisible by 8

9: Sum of digits $1 + 2 + 3 = 6$ → NOT divisible by 9

10: Last digit is 0 → divisible by 10

Answer: $2, 3, 4, 5, 6, 8, 10$

b 156

2: Last digit is even → divisible by 2

3: Sum of digits $1 + 5 + 6 = 12$ → divisible by 3

4: Last 2 is digits 56 $(56 \div 4 = 14)$ → is divisible by 4

5: Last digit is 6 → NOT divisible by 5

6: Even and divisible by 3 → divisible by 6

7: $15 - (2 \times 6) = 3$ → NOT divisible by 7

8: Last three digits $120 \div 8 = 19.5$ → NOT divisible by 8

9: Sum of digits $1 + 5 + 6 = 12$ → NOT divisible by 9

10: Last digit is 6 → NOT divisible by 10

Answer: $2, 3, 4, 6$

c 198

2: Last digit is even → divisible by 2

3: Sum of digits $1 + 9 + 8 = 18$ → divisible by 3

4: Last 2 is digits 98 $(98 \div 4 = 24.5)$ → NOT divisible by 4

5: Last digit is 8 → NOT divisible by 5

6: Even and divisible by 3 → divisible by 6

7: $19 - (2 \times 8) = 3$ → NOT divisible by 7

8: Last three digits $198 \div 8 = 24.75$ → NOT divisible by 8

9: Sum of digits $1 + 9 + 8 = 18$ → divisible by 9

10: Last digit is 8 → NOT divisible by 10

Answer: $2, 3, 6, 9$

d 240

2, 5 & 10: Last digit is 0→ divisible by 2, 5 & 10

3: Sum of digits $2 + 4 + 0 = 6$ → divisible by 3

4: Last 2 is digits 40 $(40 \div 4 = 10)$ → is divisible by 4

6: Even and divisible by 3 → divisible by 6

7: $24 - (2 \times 0) = 24$ → NOT divisible by 7

8: Last three digits $120 \div 8 = 30$ → divisible by 8

9: Sum of digits $2 + 4 + 0 = 6$ → NOT divisible by 9

Answer: $2, 3, 4, 5, 6, 8, 10$

e 357

2: Last digit is odd → NOT divisible by 2

3: Sum of digits $3 + 5 + 7 = 15$ → divisible by 3

4: Last 2 is digits 57 $(57 \div 4 = 14.25)$ → NOT divisible by 4

5 & 10: Last digit is 7 → NOT divisible by 5 or 10

6: Last digit is odd → NOT divisible by 6

7: $35 - (2 \times 7) = 21$ → divisible by 7

8: Last three digits $357 \div 8 = 44.625$ → NOT divisible by 8

9: Sum of digits $3 + 5 + 7 = 15$ → NOT divisible by 9

Answer: $3, 7$

f 378

2: Last digit is even → divisible by 2

3: Sum of digits $3 + 7 + 8 = 18$ → divisible by 3

4: Last 2 is digits 78 $(78 \div 4 = 19.5)$ → NOT divisible by 4

5: Last digit is 8 → NOT divisible by 5

6: Even and divisible by 3 → divisible by 6

7: $37 - (2 \times 8) = 21$ → divisible by 7

8: Last three digits $378 \div 8 = 47.25$ → NOT divisible by 8

9: Sum of digits $3 + 7 + 8 = 18$ → divisible by 9

10: Last digit is 8 → NOT divisible by 10

Answer: $2, 3, 6, 7, 9$

g 434

2: Last digit is even → divisible by 2

3: Sum of digits $4 + 3 + 4 = 11$ → NOT divisible by 3

4: Last 2 is digits 34 $(34 \div 4 = 8.5)$ → NOT divisible by 4

5: Last digit is 8 → NOT divisible by 5

6: Not divisible by 3 → NOT divisible by 6

7: $43 - (2 \times 4) = 35$ → divisible by 7

8: Last three digits $378 \div 8 = 54.25$ → NOT divisible by 8

9: Sum of digits $4 + 3 + 4 = 11$ → NOT divisible by 9

10: Last digit is 4 → NOT divisible by 10

Answer: $2, 7$

h 480

2: Last digit is even → divisible by 2

3: Sum of digits $4 + 8 + 0 = 12$ → divisible by 3

4: Last 2 is digits 80 $(80 \div 4 = 20)$ → divisible by 4

5: Last digit is 0 → divisible by 5

6: Even and divisible by 3 → divisible by 6

7: $48 - (2 \times 0) = 48$ → NOT divisible by 7

8: Last three digits $480 \div 8 = 60$ → divisible by 8

9: Sum of digits $4 + 8 + 0 = 12$ → NOT divisible by 9

10: Last digit is 0 → divisible by 10

Answer: $2, 3, 4, 5, 6, 8, 10$

i 496

2: Last digit is even → divisible by 2

3: Sum of digits $4 + 9 + 6 = 19$ → NOT divisible by 3

4: Last 2 is digits 96 $(96 \div 4 = 24)$ → divisible by 4

5: Last digit is 6 → NOT divisible by 5

6: Not divisible by 3 → NOT divisible by 6

7: $49 - (2 \times 6) = 37$ → NOT divisible by 7

8: Last three digits $378 \div 8 = 62$ → divisible by 8

9: Sum of digits $4 + 9 + 6 = 19$ → NOT divisible by 9

10: Last digit is 6 → NOT divisible by 10

Answer: $2, 4, 8$

j 504

2: Last digit is even → divisible by 2

3: Sum of digits $5 + 0 + 4 = 9$ → divisible by 3

4: Last 2 is digits 80 $(4 \div 4 = 1)$ → divisible by 4

5: Last digit is 4 → NOT divisible by 5

6: Even and divisible by 3 → divisible by 6

7: $50 - (2 \times 4) = 42$ → divisible by 7

8: Last three digits $504 \div 8 = 63$ → divisible by 8

9: Sum of digits $5 + 0 + 4 = 9$ → divisible by 9

10: Last digit is 4 → NOT divisible by 10

Answer: $2, 3, 4, 6, 7, 8, 9$

k 739

2, 4, 5, 6, 8, 10: Last digit is odd → NOT divisible by 2, 4, 5, 6, 8, 10

3: Sum of digits $7 + 3 + 9 = 19$ → NOT divisible by 3

7: $73 - (2 \times 9) = 55 \rightarrow$ NOT divisible by 7

9: Sum of digits $7 + 3 + 9 = 19$ → NOT divisible by 9

Answer: Not divisible by any number from 2 to 10

l 840

2, 5 & 10: Last digit is 0→ divisible by 2, 5 and 10

3: Sum of digits $8 + 4 + 0 = 12$ → divisible by 3

4: Last 2 is digits 40 $(40 \div 4 = 10)$ → divisible by 4

6: Even and divisible by 3 → divisible by 6

7: $84 - (2 \times 0) = 84 \rightarrow$ divisible by 7

8: Last three digits $504 \div 8 = 105$ → divisible by 8

9: Sum of digits $8 + 4 + 0 = 12$ → NOT divisible by 9

Answer: $2, 3, 4, 5, 6, 7, 8, 10$

Challenge Yourself

1 Sum of digits must be divisible by 3 but not 9.

$3+1+8=12$

$3+4+8=15$

$328, 338, 358, 368, 388$ and 398 are NOT divisible by 3.

378 is divisible by 9.

318 and 348 are divisible by 3 but not 9 $\therefore$ the missing digits are 1 or 4.

2 **Divisibility by 4**: The last two digits must be divisible by 4. Possible options are 20, 24, or 28.

Divisibility by 7: Check which of these numbers $720, 724, 728$ is divisible by 7.

$720: 72-(2\times 0)=72$ → NOT divisible by 7

$724: 72-(2\times 4)=64$ → NOT divisible by 7

$728: 72-(2\times 8)=56$ → divisible by 7

728 is divisible by 4 and 7 $\therefore$ the missing digit is 8.

3 **Divisibility by 2 and 5**: The last digit is 0, so any first digit will make number divisible by 2 and 5.

Divisibility by 7: Test each digit from 1-9 to find the first digit that makes the number divisible by 7.

$150:\ 15-(2\times 0)=15$ → NOT divisible by 7

$250:\ 25-(2\times 0)=25$ → divisible by 7

$350:\ 35-(2\times 0)=35$ → divisible by 7

350 is divisible by $2, 5$ and 7 $\therefore$ the smallest possible missing digit is 3.

4 **Divisibility by 3 and 9:** The sum of the digits must be divisible by 3 and 9.

$6 + 0 + 3 = 9$

$6 + 9 + 3 = 18$

$613, 623, 633, 643, 653, 663, 673$ and 683 are NOT divisible by both 3 and 9.

Divisibility by 7: Check which of these numbers 603, or 693 is divisible by 7.

$603\colon\ 60 - (2 \times 3) = 54 \rightarrow$ NOT divisible by 7

$693\colon\ 69 - (2 \times 3) = 63 \rightarrow$ divisible by 7

693 is divisible by $3, 9$ and $7 \therefore$ the missing digit is 9.

5 **Divisibility by 6**: The last digit is even so number must be divisible by 3.
Divisibility by 3: The sum of digits must be divisible by 3.

$2 + 0 + 4 = 6$

$5 + 0 + 4 = 9$

$8 + 0 + 4 = 12$

$204, 504$ and 804 are divisible by 2 and $3 \therefore$ divisible by 6.

$304, 404, 604, 704, 904$ are NOT divisible by 3.

Divisibility by 7:

$204\colon\ 20 - (2 \times 4) = 12 \rightarrow$ NOT divisible by 7

$504\colon\ 50 - (2 \times 4) = 42 \rightarrow$ divisible by 7

$804\colon\ 80 - (2 \times 4) = 72 \rightarrow$ NOT divisible by 7

504 is also divisible by 7 $\therefore$ the missing digit is 5 (as no other option is left)

Divisibility by 8: Use long division to double check

$504 \div 8 = 63$

Simplifying Fractions

Multiple Choice

1 **Option B**

$$\frac{9}{18}$$

$$= \frac{1 \times 9}{2 \times 9}$$

$$= \frac{1}{2}$$

2 **Option A**

$$\frac{12}{15}$$

$$= \frac{3 \times 4}{3 \times 5}$$

$$= \frac{4}{5}$$

3 **Option C**

$$\frac{6}{54}$$

$$= \frac{6 \times 1}{6 \times 9}$$

$$= \frac{1}{9}$$

4 **Option B**

$$\frac{42}{48}$$

$$= \frac{7 \times 6}{8 \times 6}$$

$$= \frac{7}{8}$$

5 **Option D**

$$\frac{12}{16} = \frac{3 \times 4}{4 \times 4} = \frac{3}{4}$$

$$\frac{21}{28} = \frac{7 \times 3}{7 \times 4} = \frac{3}{4}$$

$$\frac{33}{44} = \frac{11 \times 3}{11 \times 4} = \frac{3}{4}$$

$$\frac{20}{24} = \frac{4 \times 5}{4 \times 6} = \frac{5}{6}$$

Time To Practice

1 a $\frac{4}{36}$
$= \frac{4 \times 1}{4 \times 9}$
$= \frac{1}{9}$

b $\frac{5}{35}$
$= \frac{5 \times 1}{5 \times 7}$
$= \frac{1}{7}$

c $\frac{6}{18}$
$= \frac{6 \times 1}{6 \times 3}$
$= \frac{1}{3}$

2 a $\frac{9}{63}$
$= \frac{9 \times 1}{9 \times 7}$
$= \frac{1}{7}$

b $\frac{8}{56}$
$= \frac{8 \times 1}{8 \times 8}$
$= \frac{1}{8}$

c $\frac{7}{77}$
$= \frac{7 \times 1}{7 \times 11}$
$= \frac{1}{11}$

3 a $\frac{14}{21}$
$= \frac{7 \times 2}{7 \times 3}$
$= \frac{2}{3}$

b $\frac{16}{40}$
$= \frac{8 \times 2}{8 \times 5}$
$= \frac{2}{5}$

c $\frac{18}{81}$
$= \frac{9 \times 2}{9 \times 9}$
$= \frac{2}{9}$

4 a $\frac{21}{35}$
$= \frac{7 \times 3}{7 \times 5}$
$= \frac{3}{5}$

b $\frac{22}{55}$
$= \frac{11 \times 2}{11 \times 5}$
$= \frac{2}{5}$

c $\frac{24}{36}$
$= \frac{12 \times 2}{12 \times 3}$
$= \frac{2}{3}$

5 **a** $\frac{28}{63}$

$= \frac{7 \times 4}{7 \times 9}$

$= \frac{4}{9}$

b $\frac{30}{35}$

$= \frac{5 \times 6}{5 \times 7}$

$= \frac{6}{7}$

c $\frac{49}{56}$

$= \frac{7 \times 7}{7 \times 8}$

$= \frac{7}{8}$

6 **a** $\frac{50}{80}$

$= \frac{10 \times 5}{10 \times 8}$

$= \frac{5}{8}$

b $\frac{55}{99}$

$= \frac{11 \times 5}{11 \times 9}$

$= \frac{5}{9}$

c $\frac{60}{72}$

$= \frac{12 \times 5}{12 \times 6}$

$= \frac{5}{6}$

7 **a** $\frac{72}{99}$

$= \frac{9 \times 8}{9 \times 11}$

$= \frac{8}{11}$

b $\frac{84}{108}$

$= \frac{12 \times 7}{12 \times 9}$

$= \frac{7}{9}$

c $\frac{132}{144}$

$= \frac{12 \times 11}{12 \times 12}$

$= \frac{11}{12}$

Challenge Yourself

1 **a** $\frac{80}{120}$

$= \frac{10 \times 8}{10 \times 12}$

$= \frac{8}{12}$

$= \frac{4 \times 2}{4 \times 3}$

$= \frac{2}{3}$

b $\frac{54}{72}$

$= \frac{9 \times 6}{9 \times 8}$

$= \frac{6}{8}$

$= \frac{2 \times 3}{2 \times 4}$

$= \frac{3}{4}$

c $\frac{88}{110}$

$= \frac{11 \times 8}{11 \times 10}$

$= \frac{8}{10}$

$= \frac{2 \times 4}{2 \times 5}$

$= \frac{4}{5}$

2 **a** $\frac{120}{144}$

$= \frac{12 \times 10}{12 \times 12}$

$= \frac{10}{12}$

$= \frac{2 \times 5}{2 \times 6}$

$= \frac{5}{6}$

b $\frac{125}{175}$

$= \frac{5 \times 25}{5 \times 35}$

$= \frac{25}{35}$

$= \frac{5 \times 5}{5 \times 7}$

$= \frac{5}{7}$

c $\frac{300}{330}$

$= \frac{10 \times 30}{10 \times 33}$

$= \frac{30}{33}$

$= \frac{3 \times 10}{3 \times 11}$

$= \frac{10}{11}$

Adding and Subtracting Fractions

Multiple Choice

1 **Option D**

$$\frac{4}{5}-\frac{2}{3}$$

$$=\frac{12-10}{15}$$

$$=\frac{2}{15}$$

2 **Option A**

$$\frac{11}{12}-\frac{5}{6}=\frac{11}{12}-\frac{5\times 2}{6\times 2}$$

$$=\frac{11}{12}-\frac{10}{12}$$

$$=\frac{1}{12}$$

3 **Option B**

$$\frac{7}{10}-1$$

$$=\frac{7}{10}-\frac{10}{10}$$

$$=-\frac{3}{10}$$

4 **Option B**

$$1+\left(\frac{3}{5}-\frac{2}{7}\right)=1+\left(\frac{21-10}{35}\right)$$

$$=1+\frac{11}{35}$$

$$=\frac{35}{35}+\frac{11}{35}$$

$$=\frac{46}{35}$$

5 **Option C**

$$\frac{2}{3}-\left(\frac{5}{6}-\frac{1}{4}\right)=\frac{2}{3}-\left(\frac{20-6}{24}\right)=\frac{2}{3}-\frac{14}{24}$$

$$=\frac{2}{3}-\frac{7}{12}=\frac{2\times 4}{3\times 4}-\frac{7}{12}$$

$$=\frac{8}{12}-\frac{7}{12}$$

$$=\frac{1}{12}$$

Time To Practice

1

a $\frac{1}{4}+\frac{2}{4}$

$=\frac{3}{4}$

b $\frac{3}{8}-\frac{1}{8}$

$=\frac{2}{8}=\frac{1}{4}$

c $\frac{1}{10}+\frac{9}{10}$

$=\frac{10}{10}=1$

2

a $\frac{5}{12}-\frac{1}{4}$

$=\frac{5}{12}-\frac{3}{12}$

$=\frac{2}{12}=\frac{1}{6}$

b $-\frac{7}{10}+\frac{4}{5}$

$=-\frac{7}{10}+\frac{8}{10}$

$=\frac{1}{10}$

c $\frac{2}{3}-\frac{5}{6}$

$=\frac{4}{6}-\frac{5}{6}$

$=-\frac{1}{6}$

3

a $\frac{1}{8}+\frac{3}{4}$

$=\frac{4+24}{32}$

$=\frac{28}{32}=\frac{4\times 7}{4\times 8}$

$=\frac{7}{8}$

b $\frac{4}{7}+\frac{3}{10}$

$=\frac{40+21}{70}$

$=\frac{61}{70}$

c $\frac{9}{11}-\frac{3}{8}$

$=\frac{72-33}{88}$

$=\frac{39}{88}$

4

a $2-\frac{1}{4}$

$=\frac{2}{1}-\frac{1}{4}$

$=\frac{8-1}{4}$

$=\frac{7}{4}$

b $-1+\frac{1}{6}$

$=-\frac{1}{1}+\frac{1}{6}$

$=\frac{-6+1}{6}$

$=-\frac{5}{6}$

c $3+\frac{7}{10}$

$=\frac{3}{1}+\frac{7}{10}$

$=\frac{30+7}{10}$

$=\frac{37}{10}$

5 **a**

$$1+\left(\frac{7}{24}-\frac{5}{12}\right)$$

$$=1+\left(\frac{7}{24}-\frac{10}{24}\right)$$

$$=1+\left(-\frac{3}{24}\right)$$

$$=1-\frac{1}{8}$$

$$=\frac{8}{8}-\frac{1}{8}$$

$$=\frac{7}{8}$$

b

$$-1+\left(2-\frac{1}{2}\right)$$

$$=-1+\left(\frac{2}{1}-\frac{1}{2}\right)$$

$$=-1+\left(\frac{4-1}{2}\right)$$

$$=-1+\frac{3}{2}$$

$$=-\frac{2}{2}+\frac{3}{2}$$

$$=\frac{1}{2}$$

c

$$3-\left(\frac{9}{10}+\frac{2}{5}\right)$$

$$=3-\left(\frac{9}{10}+\frac{4}{10}\right)$$

$$=3-\left(\frac{13}{10}\right)$$

$$=3-\frac{13}{10}$$

$$=\frac{3}{1}-\frac{13}{10}$$

$$=\frac{30-13}{10}$$

$$=\frac{17}{10}$$

6 **a** $\left(\frac{1}{4}+\frac{2}{5}\right)-\frac{1}{10}$

$=\left(\frac{5+8}{20}\right)-\frac{1}{10}$

$=\frac{13}{20}-\frac{1}{10}$

$=\frac{13}{20}-\frac{2}{20}$

$=\frac{11}{20}$

b $-\frac{4}{9}+\left(\frac{1}{3}-\frac{1}{12}\right)$

$=-\frac{4}{9}+\left(\frac{4}{12}-\frac{1}{12}\right)$

$=-\frac{4}{9}+\frac{3}{12}$

$=\frac{-48+27}{108}$

$=-\frac{21}{108}$

$=-\frac{3\times 7}{3\times 36}$

$=-\frac{7}{36}$

c $\left(-\frac{1}{6}+\frac{1}{4}\right)+\frac{5}{8}$

$=\left(\frac{-4+6}{24}\right)+\frac{5}{8}$

$=\frac{2}{24}+\frac{5}{8}$

$=\frac{2}{24}+\frac{5}{8}$

$=\frac{2}{24}+\frac{15}{24}$

$=\frac{17}{24}$

7 **a**

$$\left(-\frac{7}{10}-\frac{2}{3}\right)-\frac{1}{5}$$

$$=\left(\frac{-21-20}{30}\right)-\frac{1}{5}$$

$$=-\frac{41}{30}-\frac{1}{5}$$

$$=-\frac{41}{30}-\frac{6}{30}$$

$$=-\frac{47}{30}$$

b

$$\frac{3}{5}+\left(-\frac{3}{4}-\frac{1}{10}\right)$$

$$=\frac{3}{5}+\left(\frac{-30-4}{40}\right)$$

$$=\frac{3}{5}+\left(-\frac{34}{40}\right)$$

$$=\frac{3}{5}-\frac{34}{40}$$

$$=\frac{3}{5}-\frac{17}{20}$$

$$=\frac{12}{20}-\frac{17}{20}$$

$$=-\frac{5}{20}$$

$$=-\frac{1}{4}$$

c

$$\left(-\frac{2}{3}-\frac{4}{5}\right)+\frac{2}{3}$$

$$=\left(\frac{-10-12}{15}\right)+\frac{2}{3}$$

$$=-\frac{22}{15}+\frac{2}{3}$$

$$=-\frac{22}{15}+\frac{10}{15}$$

$$=-\frac{12}{15}=-\frac{3\times 4}{3\times 5}$$

$$=-\frac{4}{5}$$

Challenge Yourself

1 a $\left(\frac{8}{13}+\frac{1}{2}\right)-\frac{3}{26}$

$=\left(\frac{16+13}{26}\right)-\frac{3}{26}$

$=\frac{29}{26}-\frac{3}{26}$

$=\frac{26}{26}$

$=1$

b $-\frac{11}{20}+\left(2-\frac{1}{2}\right)$

$=-\frac{11}{20}+\left(\frac{2}{1}-\frac{1}{2}\right)$

$=-\frac{11}{20}+\left(\frac{4-1}{2}\right)$

$=-\frac{11}{20}+\frac{3}{2}$

$=-\frac{11}{20}+\frac{30}{20}$

$=\frac{19}{20}$

2 a $\frac{1}{3}-\left(\frac{5}{6}-\frac{7}{18}\right)$

$=\frac{1}{3}-\left(\frac{15}{18}-\frac{7}{18}\right)$

$=\frac{1}{3}-\frac{8}{18}$

$=\frac{1}{3}-\frac{4}{9}$

$=\frac{3}{9}-\frac{4}{9}$

$=-\frac{1}{9}$

b $-\frac{2}{5}-\left(\frac{1}{10}-\frac{1}{15}\right)$

$=-\frac{2}{5}-\left(\frac{15-10}{150}\right)$

$=-\frac{2}{5}-\frac{5}{150}$

$=-\frac{2}{5}-\frac{1}{30}$

$=-\frac{12}{30}-\frac{1}{30}$

$=-\frac{13}{30}$

3 **a** $\frac{2}{18}-\left(\frac{2}{9}-\frac{4}{27}\right)$

$=\frac{1}{9}-\left(\frac{6}{27}-\frac{4}{27}\right)$

$=\frac{1}{9}-\frac{2}{27}$

$=\frac{3}{27}-\frac{2}{27}$

$=\frac{1}{27}$

b $\left(\frac{5}{8}-\frac{1}{10}\right)-\frac{3}{4}$

$=\left(\frac{50-8}{80}\right)-\frac{3}{4}$

$=\frac{42}{80}-\frac{3}{4}$

$=\frac{21}{40}-\frac{3}{4}$

$=\frac{21}{40}-\frac{30}{40}$

$=-\frac{9}{40}$

Multiplying and Dividing Fractions

Multiple Choice

1 **Option B**

$$\frac{2}{5} \times \frac{3}{4}$$

$$= \frac{6}{20}$$

$$= \frac{3}{10}$$

2 **Option D**

$$\frac{5}{6} \div \frac{10}{9} = \frac{5}{6} \times \frac{9}{10}$$

$$= \frac{1}{2} \times \frac{3}{2}$$

$$= \frac{3}{4}$$

3 **Option C**

$$6 \div \frac{2}{5} = 6 \times \frac{5}{2}$$

$$= \frac{30}{2}$$

$$= 15$$

4 **Option B**

$$\frac{9}{14} \times \frac{7}{3}$$

$$= \frac{3 \times 3}{2 \times 7} \times \frac{7}{3}$$

$$= \frac{3}{2} \times \frac{1}{1}$$

$$= \frac{3}{2}$$

5 **Option A**

$$\frac{5}{8} \div 4 = \frac{5}{8} \div \frac{4}{1}$$

$$= \frac{5}{8} \times \frac{1}{4}$$

$$= \frac{5}{32}$$

Time To Practice

1 a $\frac{2}{9} \times \frac{1}{5}$

$= \frac{2}{45}$

b $\frac{1}{5} \times \frac{3}{2}$

$= \frac{3}{10}$

c $\frac{7}{8} \times \frac{9}{11}$

$= \frac{63}{88}$

2 a $\frac{2}{3} \times 4$

$= \frac{8}{3}$

b $\frac{2}{3} \div 4$

$= \frac{2}{3} \times \frac{1}{4}$

$= \frac{2}{12}$

$= \frac{1}{6}$

c $4 \div \frac{2}{3}$

$= 4 \times \frac{3}{2}$

$= \frac{12}{2}$

$= 6$

3 a $\frac{5}{6} \times \frac{18}{25}$

$= \frac{1}{1} \times \frac{3}{5}$

$= \frac{3}{5}$

b $-\frac{7}{9} \times \frac{45}{49}$

$= -\frac{1}{1} \times \frac{5}{7}$

$= -\frac{5}{7}$

c $\frac{3}{40} \times \frac{5}{33}$

$= \frac{1}{8} \times \frac{1}{11}$

$= \frac{1}{88}$

4 a $\frac{1}{6} \div \frac{4}{5}$

$= \frac{1}{6} \times \frac{5}{4}$

$= \frac{5}{24}$

b $\frac{2}{3} \div \frac{7}{8}$

$= \frac{2}{3} \times \frac{8}{7}$

$= \frac{16}{21}$

c $\frac{4}{9} \div \frac{5}{8}$

$= \frac{4}{9} \times \frac{8}{5}$

$= \frac{32}{45}$

5 **a** $\left(\frac{2}{3} \div \frac{1}{4}\right) \div 2$

$= \left(\frac{2}{3} \times \frac{4}{1}\right) \div 2$

$= \frac{8}{3} \times \frac{1}{2}$

$= \frac{8}{6}$

$= \frac{4}{3}$

b $\frac{5}{9} \div \left(\frac{5}{6} \div \frac{2}{3}\right)$

$= \frac{5}{9} \div \left(\frac{5}{6} \times \frac{3}{2}\right)$

$= \frac{5}{9} \div \left(\frac{5}{2} \times \frac{1}{2}\right)$

$= \frac{5}{9} \div \frac{5}{4}$

$= \frac{5}{9} \times \frac{4}{5}$

$= \frac{4}{9}$

c $\left(\frac{1}{2} \div \frac{45}{40}\right) \div \frac{40}{25}$

$= \left(\frac{1}{2} \div \frac{9}{8}\right) \div \frac{8}{5}$

$= \left(\frac{1}{2} \times \frac{8}{9}\right) \div \frac{8}{5}$

$= \left(\frac{1}{1} \times \frac{4}{9}\right) \div \frac{8}{5}$

$= \frac{4}{9} \times \frac{5}{8}$

$= \frac{1}{9} \times \frac{5}{2}$

$= \frac{5}{18}$

6 **a** $\left(\frac{1}{2} - \frac{2}{3}\right) \times \frac{3}{4}$

$= \left(\frac{3-4}{6}\right) \times \frac{3}{4}$

$= -\frac{1}{6} \times \frac{3}{4}$

$= -\frac{1}{2} \times \frac{1}{4}$

$= -\frac{1}{8}$

b $\frac{3}{5} \div \left(\frac{1}{4} + \frac{1}{6}\right)$

$= \frac{3}{5} \div \left(\frac{6+4}{24}\right)$

$= \frac{3}{5} \div \frac{10}{24}$

$= \frac{3}{5} \div \frac{5}{12}$

$= \frac{3}{5} \times \frac{12}{5}$

$= \frac{36}{25}$

c $\left(\frac{5}{8} \times \frac{2}{3}\right) - \frac{1}{4}$

$= \left(\frac{5}{4} \times \frac{1}{3}\right) - \frac{1}{4}$

$= \frac{5}{12} - \frac{1}{4}$

$= \frac{5}{12} - \frac{3}{12}$

$= \frac{2}{12}$

$= \frac{1}{6}$

7 **a** $\left(\frac{12}{25} \div \frac{8}{5}\right) \times \frac{7}{10}$

$= \left(\frac{12}{25} \times \frac{5}{8}\right) \times \frac{7}{10}$

$= \left(\frac{3}{5} \times \frac{1}{2}\right) \times \frac{7}{10}$

$= \frac{3}{10} \times \frac{7}{10}$

$= \frac{21}{100}$

b $\left(\frac{55}{30} \div \frac{40}{30}\right) \times \frac{10}{11}$

$= \left(\frac{11}{6} \div \frac{4}{3}\right) \times \frac{10}{11}$

$= \left(\frac{11}{6} \times \frac{3}{4}\right) \times \frac{10}{11}$

$= \frac{33}{24} \times \frac{10}{11}$

$= \frac{3}{12} \times \frac{5}{1}$

$= \frac{1}{4} \times \frac{5}{1}$

$= \frac{5}{4}$

c $\frac{5}{2} \times \left(\frac{160}{30} \div \frac{2}{15}\right)$

$= \frac{5}{2} \times \left(\frac{16}{3} \times \frac{15}{2}\right)$

$= \frac{5}{2} \times \left(\frac{8}{1} \times \frac{5}{1}\right)$

$= \frac{5}{2} \times \frac{40}{1}$

$= \frac{5}{1} \times \frac{20}{1}$

$= 100$

Challenge Yourself

1 a

$$\left(\frac{8}{9} \div \frac{2}{3}\right) \div \left(\frac{3}{4} \div \frac{1}{2}\right)$$

$$= \left(\frac{8}{9} \times \frac{3}{2}\right) \div \left(\frac{3}{4} \times \frac{2}{1}\right)$$

$$= \left(\frac{4}{3} \times \frac{1}{1}\right) \div \left(\frac{3}{2} \times \frac{1}{1}\right)$$

$$= \frac{4}{3} \div \frac{3}{2}$$

$$= \frac{4}{3} \times \frac{2}{3}$$

$$= \frac{8}{9}$$

b

$$\left(\frac{11}{12} \div \frac{5}{6}\right) \div \left(\frac{3}{14} \div \frac{1}{4}\right)$$

$$= \left(\frac{11}{12} \times \frac{6}{5}\right) \div \left(\frac{3}{14} \times \frac{4}{1}\right)$$

$$= \left(\frac{11}{2} \times \frac{1}{5}\right) \div \left(\frac{3}{7} \times \frac{2}{1}\right)$$

$$= \frac{11}{10} \div \frac{6}{7}$$

$$= \frac{11}{10} \times \frac{7}{6}$$

$$= \frac{77}{60}$$

2 a

$$\left(\frac{5}{8} \div \frac{1}{4}\right) \times \left(\frac{3}{5} - \frac{1}{10}\right)$$

$$= \left(\frac{5}{8} \times \frac{4}{1}\right) \times \left(\frac{30 - 5}{50}\right)$$

$$= \left(\frac{5}{2} \times \frac{1}{1}\right) \times \left(\frac{25}{50}\right)$$

$$= \frac{5}{2} \times \frac{1}{2}$$

$$= \frac{5}{4}$$

b

$$\frac{4}{5} \times \left(\frac{2}{3} + \frac{1}{6}\right) \div \frac{3}{4}$$

$$= \frac{4}{5} \times \left(\frac{4}{6} + \frac{1}{6}\right) \div \frac{3}{4}$$

$$= \frac{4}{5} \times \frac{5}{6} \div \frac{3}{4}$$

$$= \frac{2}{1} \times \frac{1}{3} \div \frac{3}{4}$$

$$= \frac{2}{3} \times \frac{4}{3}$$

$$= \frac{8}{9}$$

3 **a** $\left(\frac{18}{25} \div \frac{12}{35}\right) + \left(\frac{16}{45} \times \frac{27}{32}\right)$

$= \left(\frac{18}{25} \times \frac{35}{12}\right) + \left(\frac{1}{5} \times \frac{3}{2}\right)$

$= \left(\frac{3}{5} \times \frac{7}{2}\right) + \frac{3}{10}$

$= \frac{21}{10} + \frac{3}{10}$

$= \frac{24}{10}$

$= \frac{12}{5}$

b $\left(\frac{24}{35} \times \frac{15}{28}\right) - \left(\frac{3}{14} \div \frac{21}{2}\right)$

$= \left(\frac{6}{7} \times \frac{3}{7}\right) - \left(\frac{3}{14} \times \frac{2}{21}\right)$

$= \frac{18}{49} - \left(\frac{1}{7} \times \frac{1}{7}\right)$

$= \frac{18}{49} - \frac{1}{49}$

$= \frac{17}{49}$

Decimals to Fractions

Multiple Choice

1 **Option C**

$$\frac{0.45}{1} \times \frac{100}{100}$$

$$= \frac{45}{100} = \frac{5 \times 9}{5 \times 20}$$

$$= \frac{9}{20}$$

2 **Option B**

$$\frac{0.66}{1} \times \frac{100}{100}$$

$$= \frac{66}{100} = \frac{2 \times 33}{2 \times 50}$$

$$= \frac{33}{50}$$

3 **Option D**

$$\frac{0.7}{1} \times \frac{10}{100} = \frac{7}{10}$$

$$\frac{21}{30} = \frac{3 \times 7}{3 \times 10} = \frac{7}{10}$$

$$\frac{3}{4} \neq 0.7$$

4 **Option B**

$$\frac{1.5}{1} \times \frac{10}{10}$$

$$= \frac{15}{10} = \frac{5 \times 3}{5 \times 2}$$

$$= \frac{3}{2}$$

5 **Option A**

$$\frac{2.25}{1} \times \frac{100}{100} = \frac{225}{100}$$

$$= \frac{25 \times 9}{25 \times 4}$$

$$= \frac{9}{4}$$

Time To Practice

1 a 0.3

$$= \frac{0.3}{1} \times \frac{10}{10}$$

$$= \frac{3}{10}$$

b 0.7

$$= \frac{0.7}{1} \times \frac{10}{10}$$

$$= \frac{7}{10}$$

c 0.8

$$= \frac{0.8}{1} \times \frac{10}{10}$$

$$= \frac{8}{10}$$

$$= \frac{4}{5}$$

d 0.12

$$= \frac{0.12}{1} \times \frac{100}{100}$$

$$= \frac{12}{100}$$

$$= \frac{4 \times 3}{4 \times 25}$$

$$= \frac{3}{25}$$

e 0.13

$$= \frac{0.13}{1} \times \frac{100}{100}$$

$$= \frac{13}{100}$$

f 0.22

$$= \frac{0.22}{1} \times \frac{100}{100}$$

$$= \frac{22}{100}$$

$$= \frac{2 \times 11}{2 \times 50}$$

$$= \frac{11}{50}$$

g 0.31

$$= \frac{0.31}{1} \times \frac{100}{100}$$

$$= \frac{31}{100}$$

h 0.42

$$= \frac{0.42}{1} \times \frac{100}{100}$$

$$= \frac{42}{100}$$

$$= \frac{2 \times 21}{2 \times 50}$$

$$= \frac{21}{50}$$

i 0.55

$$= \frac{0.55}{1} \times \frac{100}{100}$$

$$= \frac{55}{100}$$

$$= \frac{5 \times 11}{5 \times 20}$$

$$= \frac{11}{20}$$

1 j 0.62

$= \frac{0.62}{1} \times \frac{100}{100}$

$= \frac{62}{100}$

$= \frac{2 \times 31}{2 \times 50}$

$= \frac{31}{50}$

k 0.78

$= \frac{0.78}{1} \times \frac{100}{100}$

$= \frac{78}{100}$

$= \frac{2 \times 39}{2 \times 50}$

$= \frac{39}{50}$

l 0.82

$= \frac{0.82}{1} \times \frac{100}{100}$

$= \frac{82}{100}$

$= \frac{2 \times 41}{2 \times 50}$

$= \frac{41}{50}$

m 0.88

$= \frac{0.88}{1} \times \frac{100}{100}$

$= \frac{88}{100}$

$= \frac{4 \times 22}{4 \times 25}$

$= \frac{22}{25}$

n 0.92

$= \frac{0.92}{1} \times \frac{100}{100}$

$= \frac{92}{100}$

$= \frac{4 \times 23}{4 \times 25}$

$= \frac{23}{25}$

o 0.95

$= \frac{0.95}{1} \times \frac{100}{100}$

$= \frac{95}{100}$

$= \frac{5 \times 19}{5 \times 20}$

$= \frac{19}{20}$

2 **a** 1.3

$$= \frac{1.3}{1} \times \frac{10}{10}$$

$$= \frac{13}{10}$$

b 1.4

$$= \frac{1.4}{1} \times \frac{10}{10}$$

$$= \frac{14}{10}$$

$$= \frac{2 \times 7}{2 \times 5}$$

$$= \frac{7}{5}$$

c 1.6

$$= \frac{1.6}{1} \times \frac{10}{10}$$

$$= \frac{16}{10}$$

$$= \frac{2 \times 8}{2 \times 5}$$

$$= \frac{8}{5}$$

d 1.9

$$= \frac{1.9}{1} \times \frac{10}{10}$$

$$= \frac{19}{10}$$

e 2.5

$$= \frac{2.5}{1} \times \frac{10}{10}$$

$$= \frac{25}{10}$$

$$= \frac{5 \times 5}{5 \times 2}$$

$$= \frac{5}{2}$$

f 3.9

$$= \frac{3.9}{1} \times \frac{10}{10}$$

$$= \frac{39}{10}$$

g 1.46

$$= \frac{1.46}{1} \times \frac{100}{100}$$

$$= \frac{146}{100}$$

$$= \frac{2 \times 73}{2 \times 50}$$

$$= \frac{73}{50}$$

h 1.72

$$= \frac{1.72}{1} \times \frac{100}{100}$$

$$= \frac{172}{100}$$

$$= \frac{4 \times 43}{4 \times 25}$$

$$= \frac{43}{25}$$

i 1.85

$$= \frac{1.85}{1} \times \frac{100}{100}$$

$$= \frac{185}{100}$$

$$= \frac{5 \times 37}{5 \times 20}$$

$$= \frac{37}{20}$$

Challenge Yourself

1 a 0.03

$$= \frac{0.03}{1} \times \frac{100}{100}$$

$$= \frac{3}{100}$$

b 0.04

$$= \frac{0.04}{1} \times \frac{100}{100}$$

$$= \frac{4}{100}$$

$$= \frac{4 \times 1}{4 \times 25}$$

$$= \frac{1}{25}$$

c 0.005

$$= \frac{0.005}{1} \times \frac{1000}{1000}$$

$$= \frac{5}{1000}$$

$$= \frac{5 \times 1}{5 \times 200}$$

$$= \frac{1}{200}$$

d 0.123

$$= \frac{0.123}{1} \times \frac{1000}{1000}$$

$$= \frac{123}{1000}$$

e 0.402

$$= \frac{0.402}{1} \times \frac{1000}{1000}$$

$$= \frac{402}{1000}$$

$$= \frac{2 \times 201}{2 \times 500}$$

$$= \frac{201}{500}$$

f 0.888

$$= \frac{0.888}{1} \times \frac{1000}{1000}$$

$$= \frac{888}{1000}$$

$$= \frac{8 \times 111}{8 \times 125}$$

$$= \frac{111}{125}$$

Exponents
Powers

Multiplying and Dividing Powers

Multiple Choice

1 **Option D**

$x^3 \times x^4$

$= x^{3+4}$

$= x^7$

2 **Option B**

$a^5 \div a^2$

$= a^{5-2}$

$= a^3$

3 **Option A**

$3y^2 \times 2y^5$

$= (3 \times 2)y^{2+5}$

$= 6y^7$

4 **Option C**

$\frac{p^4 \times p^2}{p^3}$

$= \frac{p^{4+2}}{p^3}$

$= \frac{p^6}{p^3}$

$= p^{6-3}$

$= p^3$

5 **Option B**

$x^0 \times x^4$

$= 1 \times x^4$

$= x^4$

Time To Practice

1 a $2^3 \times 2^2$
$= 2^{3+2}$
$= 2^5 = 32$

b $3^2 \times 3$
$= 3^{2+1}$
$= 3^3 = 27$

c $5^0 \times 5^3$
$= 1 \times 5^3$
$= 5^3 = 125$

d $x^6 \times x^0$
$= x^6 \times 1$
$= x^6$

e $y^3 \times y^2$
$= y^{3+2}$
$= y^5$

f $z^9 \times z^7$
$= z^{9+7}$
$= z^{16}$

g $a^3 \times a^4 \times a^6$
$= a^{3+4+6}$
$= a^{13}$

h $b^3 \times b^5 \times b^2$
$= b^{3+5+2}$
$= b^{10}$

i $c^2 \times c^8 \times c^9$
$= c^{2+8+9}$
$= c^{19}$

j $5 \times (2x^2)^0$
$= 5 \times 1$
$= 5$

k $(3x^2y)^0 \times 7$
$= 1 \times 7$
$= 7$

l $3(2x^2y^3)^0 - 4$
$= 3 \times 1 - 4$
$= -1$

m $6a^4 \times 2a^2$
$= 12a^{4+2}$
$= 12a^6$

m $10b^6 \times 2b^3$
$= 20b^{6+3}$
$= 20b^9$

o $5c^0 \times 7c^5 \times c^2$
$= 35c^{5+2}$
$= 35c^7$

2 a $7^6 \div 7^4$
$= 7^{6-4}$
$= 7^2 = 49$

b $4^{12} \div 4^9$
$= 4^{12-9}$
$= 4^3 = 64$

c $10^{17} \div 10^{14}$
$= 10^{17-14}$
$= 10^3 = 1000$

d $m^9 \div m^2$
$= m^{9-2}$
$= m^7$

e $n^{20} \div n^5$
$= n^{20-5}$
$= n^{15}$

f $p^{15} \div p^3$
$= p^{15-3}$
$= p^{12}$

2 **g** $\frac{5^5}{5^4} \times 5$

$= 5^{5-4} \times 5$

$= 5 \times 5$

$= 25$

h $\frac{3^8}{3^6} \times 3^2$

$= 3^{8-6} \times 3^2$

$= 3^2 \times 3^2$

$= 3^{2+2} = 3^4$

$= 81$

i $\frac{2^8}{2^5} \times 2^3$

$= 2^{8-5} \times 2^3$

$= 2^3 \times 2^3$

$= 2^{3+3} = 2^6$

$= 64$

j $\frac{10x^5}{2x^2} \times 3x^3$

$= 5x^{5-2} \times 3x^3$

$= 5x^3 \times 3x^3$

$= 15x^{3+3}$

$= 15x^6$

k $\frac{14y^8 \times y^0}{7y^4}$

$= \frac{14y^8 \times 1}{7y^4}$

$= 2y^{8-4}$

$= 2y^4$

l $\frac{50z^6 \times z^6}{5z^2}$

$= \frac{50z^{6+6}}{5z^2}$

$= 10z^{12-2}$

$= 10z^{10}$

m $4a^3 \times 3a^4 \div 2a^2$

$= 12a^{3+4} \div 2a^2$

$= 12a^7 \div 2a^2$

$= 6a^{7-2}$

$= 6a^5$

n $12b^9 \times 3b^2 \div 4b^0$

$= 36b^{9+2} \div 4$

$= 36b^{11} \div 4$

$= 9b^{11}$

o $16c^7 \times 2c^3 \div 8c^2$

$= 32c^{7+3} \div 8c^2$

$= 32c^{10} \div 8c^2$

$= 4c^{10-2}$

$= 4c^8$

Challenge Yourself

1 a $\frac{12u^{11}v^9 \times 3u^2v^3}{18u^6v^5}$

$= \frac{36u^{11+2}v^{9+3}}{18u^6v^5}$

$= \frac{36u^{13}v^{12}}{18u^6v^5}$

$= 2u^{13-6}v^{12-5}$

$= 2u^7v^7$

b $\frac{5e^6f^4 \times 8e^4f^3}{10ef^5}$

$= \frac{40e^{6+4}f^{4+3}}{10ef^5}$

$= \frac{40e^{10}f^7}{10ef^5}$

$= 4e^{10-1}f^{7-5}$

$= 4e^9f^2$

2 a $\frac{-3ab^3}{ab} \div \frac{b}{a^4b^2}$

$= \frac{-3a^{1-1}b^{3-1}}{1} \div \frac{b}{a^4b^2}$

$= \frac{-3b^2}{1} \times \frac{a^4b^2}{b}$

$= \frac{-3a^4b^4}{b}$

$= -3a^4b^{4-1}$

$= -3a^4b^3$

b $\frac{16a^7b^6}{6a^3b^3} \div \frac{8a^6b^3}{27a^3b^2}$

$= \frac{16a^{7-3}b^{6-3}}{6} \div \frac{8a^{6-3}b^{3-2}}{27}$

$= \frac{8a^4b^3}{3} \div \frac{8a^3b}{27}$

$= \frac{8a^4b^3}{3} \times \frac{27}{8a^3b}$

$= \frac{a^4b^3}{1} \times \frac{9}{a^3b}$

$= \frac{9a^4b^3}{a^3b}$

$= 9a^{4-3}b^{3-1}$

$= 9ab^2$

3 **a** $\dfrac{8x^2y \times 3x^5y^2}{6x^2y^2} \times \dfrac{7x^5y^2}{16x^3y}$

$= \dfrac{24x^{2+5}y^{1+2}}{6x^2y^2} \times \dfrac{7x^5y^2}{16x^3y}$

$= \dfrac{24x^7y^3}{6x^2y^2} \times \dfrac{7x^5y^2}{16x^3y}$

$= \dfrac{4x^{7-2}y^{3-2}}{1} \times \dfrac{7x^5y^2}{16x^3y}$

$= \dfrac{4x^5y}{1} \times \dfrac{7x^5y^2}{16x^3y}$

$= \dfrac{28x^{5+5}y^{2+1}}{16x^3y}$

$= \dfrac{28x^{10}y^3}{16x^3y}$

$= \dfrac{28x^{10-3}y^{3-1}}{16}$

$= \dfrac{7x^7y^2}{4}$

b $\dfrac{16x^4y^8}{27x^4y^9} \div \dfrac{x^6y^2}{x^5y^5z^3 \times xyz}$

$= \dfrac{16x^4y^8}{27x^4y^9} \div \dfrac{x^6y^2}{x^{5+1}y^{5+1}z^{3+1}}$

$= \dfrac{16x^4y^8}{27x^4y^9} \div \dfrac{x^6y^2}{x^6y^6z^4}$

$= \dfrac{16x^4y^8}{27x^4y^9} \times \dfrac{x^6y^6z^4}{x^6y^2}$

$= \dfrac{16x^{4+6}y^{8+6}z^4}{27x^{4+6}y^{9+2}}$

$= \dfrac{16x^{10}y^{14}z^4}{27x^{10}y^{11}}$

$= \dfrac{16x^{10-10}y^{14-11}z^4}{27}$

$= \dfrac{16y^3z^4}{27}$

Power of a Power

Multiple Choice

1 **Option C**

$(2ab^3)^4 = 2^4a^4b^{3\times4}$

$= 16a^4b^{12}$

2 **Option A**

$x^2y^2 \times (-3x^2y)^2$

$= x^2y^2 \times (-3)^2x^{2\times2}y^{1\times2}$

$= x^2y^2 \times 9x^4y^2$

$= 9x^6y^4$

3 **Option A**

$\left(\frac{3x^2y}{4z}\right)^3 = \frac{3^3x^{2\times3}y^3}{4^3z^3}$

$= \frac{27x^6y^3}{64z^3}$

4 **Option D**

$\frac{6a^6b^5}{a^5b^3} \times \frac{a^4}{(2b)^2}$

$= 6ab^2 \times \frac{a^4}{4b^2}$

$= \frac{6a^5b^2}{4b^2} = \frac{3a^5}{2}$

5 **Option B**

$-\frac{3x^6y^3}{xy} \div \left(\frac{x^2y}{y^3}\right)^2 = -\frac{3x^6y^3}{xy} \div \frac{x^{2\times2}y^2}{y^{3\times2}}$

$= -\frac{3x^6y^3}{xy} \div \frac{x^4y^2}{y^6}$

$= -\frac{3x^6y^3}{xy} \times \frac{y^6}{x^4y^2}$

$= 3x^5y^2 \times \frac{y^4}{x^4}$

$= -3xy^6$

Time To Practice

1 **a** $(a^2)^5$

$= a^{2\times 5}$

$= a^{10}$

b $(b^5)^2 \times (b^2)^3$

$= b^{5\times 2} \times b^{2\times 3}$

$= b^{10} \times b^6 = b^{16}$

c $(2a^6)^3$

$= 2^3 \times a^{6\times 3}$

$= 8a^{18}$

d $(-3x^4)^3$

$= (-3)^3 \times x^{4\times 3}$

$= -27x^{12}$

e $(x^2y)^3 \times (yx^3)^2$

$= x^{2\times 3}y^3 \times y^2x^{3\times 2}$

$= x^6y^3 \times y^2x^6$

$= x^{12}y^5$

f $(xy^2)^8 \div x^4y^5$

$= x^8y^{2\times 8} \div x^4y^5$

$= x^8y^{16} \div x^4y^5$

$= x^4y^{11}$

g $(2x^2y^3)^2 \times xy^4$

$= 2^2x^{2\times 2}y^{3\times 2} \times xy^4$

$= 4x^4y^6 \times xy^4$

$= 4x^5y^{10}$

h $(2x^3y^2)^4 \times (x^2y)^3$

$= 2^4x^{3\times 4}y^{2\times 4} \times x^{2\times 3}y^3$

$= 16x^{12}y^8 \times x^6y^3$

$= 16x^{18}y^{11}$

i $(2x^2y^3)^4 \div (4x^4)^2$

$= 2^4x^{2\times 4}y^{3\times 4} \div 4^2x^{4\times 2}$

$= 16x^8y^{12} \div 16x^8$

$= y^{12}$

j $(2m^3n)^5 \div (8mn^2)^2$

$= 2^5m^{3\times 5}n^5 \div 8^2m^2n^{2\times 2}$

$= 32m^{15}n^5 \div 64m^2n^4$

$= \dfrac{m^{13}n}{2}$

k $(2p)^5q^6 \div 9p^4q^3$

$= 2^5p^5q^6 \div 9p^4q^3$

$= \dfrac{32pq^3}{9}$

l $r^6s^2 \times 2(3r)^3s^5$

$= r^6s^2 \times 2 \times 3^3r^3s^5$

$= r^6s^2 \times 54r^3s^5$

$= 54r^9s^7$

2 **a** $\left(\frac{a}{b^2}\right)^2$

$= \frac{a^2}{b^4}$

b $\left(\frac{a^2}{b^3}\right)^5$

$= \frac{a^{2\times5}}{b^{3\times5}}$

$= \frac{a^{10}}{b^{15}}$

c $\left(\frac{4a^3}{b^2}\right)^2$

$= \frac{4^2a^6}{b^{2\times2}}$

$= \frac{16a^6}{b^4}$

d $\left(\frac{x^2}{y^3}\right)^4 \times \left(\frac{x^3}{y}\right)^5$

$= \frac{x^{2\times4}}{y^{3\times4}} \times \frac{x^{3\times5}}{y^5}$

$= \frac{x^8}{y^{12}} \times \frac{x^{15}}{y^5}$

$= \frac{x^{23}}{y^{17}}$

e $\left(\frac{2x^3}{y^2}\right)^2 \times \left(\frac{x^3}{2y}\right)^4$

$= \frac{2^2x^{3\times2}}{y^{2\times2}} \times \frac{x^{3\times4}}{2^4y^4}$

$= \frac{4x^6}{y^4} \times \frac{x^{12}}{16y^4}$

$= \frac{x^6}{y^4} \times \frac{x^{12}}{4y^4}$

$= \frac{x^{18}}{4y^8}$

f $\left(\frac{3x^2}{2y^3}\right)^3 \div \frac{6x}{y^2}$

$= \frac{3^3x^{2\times3}}{2^3y^{3\times3}} \div \frac{6x}{y^2}$

$= \frac{27x^6}{8y^9} \times \frac{y^2}{6x}$

$= \frac{9x^6}{8y^9} \times \frac{y^2}{2x}$

$= \frac{9x^6y^2}{16xy^9}$

$= \frac{9x^5}{16y^7}$

2 **g** $\left(\frac{6a^4}{b^6}\right)^2 \div \left(\frac{3a}{b^2}\right)^4$

$= \frac{6^2a^{4\times2}}{b^{6\times2}} \div \frac{3^4a^4}{b^{2\times4}}$

$= \frac{36a^8}{b^{12}} \div \frac{81a^4}{b^8}$

$= \frac{36a^8}{b^{12}} \times \frac{b^8}{81a^4}$

$= \frac{4a^8}{b^{12}} \times \frac{b^8}{9a^4}$

$= \frac{4a^8b^8}{9a^4b^{12}}$

$= \frac{4a^4}{9b^4}$

h $\left(\frac{9a^4}{2b^2}\right)^2 \div \left(\frac{3b^2}{4a^3}\right)^3$

$= \frac{9^2a^{4\times2}}{2^2b^{2\times2}} \div \frac{3^3b^{2\times3}}{4^3a^{3\times3}}$

$= \frac{81a^8}{4b^4} \div \frac{27b^6}{64a^9}$

$= \frac{81a^8}{4b^4} \times \frac{64a^9}{27b^6}$

$= \frac{3a^8}{b^4} \times \frac{16a^9}{b^6}$

$= \frac{48a^{17}}{b^{10}}$

i $\left(\frac{3a^5}{4b^7}\right)^2 \div 4\left(\frac{1}{a^4b^4}\right)^7$

$= \frac{3^2a^{5\times2}}{4^2b^{7\times2}} \div 4\left(\frac{1}{a^{4\times7}b^{4\times7}}\right)$

$= \frac{9a^{10}}{16b^{14}} \div \frac{4}{a^{28}b^{28}}$

$= \frac{9a^{10}}{16b^{14}} \times \frac{a^{28}b^{28}}{4}$

$= \frac{9a^{38}b^{28}}{64b^{14}}$

$= \frac{9a^{38}b^{14}}{64}$

j $\left(\frac{3m^4}{n^3}\right)^2 \div \left(\frac{n^7}{2m^2}\right)^3$

$= \frac{3^2m^{4\times2}}{n^{3\times2}} \div \left(\frac{n^{7\times3}}{2^3m^{2\times3}}\right)$

$= \frac{9m^8}{n^6} \div \frac{n^{21}}{8m^6}$

$= \frac{9m^8}{n^6} \times \frac{8m^6}{n^{21}}$

$= \frac{72m^{14}}{n^{27}}$

2 **k** $\frac{1}{2}\left(\frac{p^2q}{q^5}\right)^3 \times \left(\frac{p}{q^8}\right)^3$

$= \frac{1}{2}\left(\frac{p^{2\times3}q^3}{q^{5\times3}}\right) \times \frac{p^3}{q^{8\times3}}$

$= \frac{p^6q^3}{2q^{15}} \times \frac{p^3}{q^{24}}$

$= \frac{p^9q^3}{2q^{39}}$

$= \frac{p^9}{2q^{36}}$

l $\frac{7}{8}\left(\frac{2r}{3r^3}\right)^3 \div \left(\frac{r^4}{s^2}\right)^3$

$= \frac{7}{8}\left(\frac{2^3r^3}{3^3r^{3\times3}}\right) \div \frac{r^{4\times3}}{s^{2\times3}}$

$= \frac{7}{8}\left(\frac{8r^3}{27r^9}\right) \div \frac{r^{12}}{s^6}$

$= \frac{7}{8}\left(\frac{8r^3}{27r^9}\right) \times \frac{s^6}{r^{12}}$

$= \frac{7}{1}\left(\frac{1}{27r^6}\right) \times \frac{s^6}{r^{12}}$

$= \frac{7}{27r^6} \times \frac{s^6}{r^{12}}$

$= \frac{7s^6}{27r^{18}}$

3 **a** $\left(\left(x^3\right)^4\right)^2$

$= x^{3\times4\times2}$

$= x^{24}$

b $\left(\left(2x^3\right)^2\right)^2$

$= 2^{1\times2\times2}x^{3\times2\times2}$

$= 2^4x^{12}$

$= 16x^{12}$

c $\left(\left(x^2y\right)^6\right)^4$

$= x^{2\times6\times4}y^{1\times6\times4}$

$= x^{48}y^{24}$

d $\left(\left(\frac{a^3}{b^2}\right)^2\right)^4$

$= \frac{a^{3\times2\times4}}{b^{2\times2\times4}}$

$= \frac{a^{24}}{b^{16}}$

3 **e** $\left(\left(\frac{a^3}{b^4c}\right)^2\right)^3$

$= \frac{a^{3\times2\times3}}{b^{4\times2\times3}c^{1\times2\times3}}$

$= \frac{a^{18}}{b^{24}c^6}$

f $\left(\left(\frac{a^9}{3b^7}\right)^2\right)^2$

$= \frac{a^{9\times2\times2}}{3^{1\times2\times2}b^{7\times2\times2}}$

$= \frac{a^{36}}{3^4b^{28}}$

$= \frac{a^{36}}{81b^{28}}$

Challenge Yourself

1 **a** $\left(\frac{4a^3b^2c^4}{2a^2bc}\right)^2 \times \left(\frac{3a^2b}{c^4}\right)^2$

$= \left(\frac{2abc^3}{1}\right)^2 \times \left(\frac{3a^2b}{c^4}\right)^2$

$= \frac{2^2a^2b^2c^6}{1} \times \frac{3^2a^4b^2}{c^8}$

$= \frac{4a^2b^2c^6}{1} \times \frac{9a^4b^2}{c^8}$

$= \frac{36a^6b^4c^6}{c^8}$

$= \frac{36a^6b^4}{c^2}$

b $\frac{\left(4a^2b^2\right)^2 \times -2(a^4b)^2}{(-3a^3b^2)^2}$

$= \frac{4^2a^4b^4 \times -2a^8b^2}{(-3)^2a^6b^4}$

$= \frac{16a^4b^4 \times -2a^8b^2}{9a^6b^4}$

$= \frac{-32a^{12}b^6}{9a^6b^4}$

$= \frac{-32a^6b^2}{9}$

2 **a** $\dfrac{(3x^4)^3}{(2y^2z^3)^2} \times \left(\dfrac{2xz}{x^3z}\right)^2$

$= \dfrac{3^3x^{12}}{2^2y^4z^6} \times \dfrac{2^2x^2z^2}{x^6z^2}$

$= \dfrac{27x^{12}}{4y^4z^6} \times \dfrac{4x^2z^2}{x^6z^2}$

$= \dfrac{27x^{12}}{y^4z^6} \times \dfrac{1}{x^4}$

$= \dfrac{27x^8}{y^4z^6}$

b $\dfrac{(-5x^3y^2)^3}{(z^2)^2} \times \left(\dfrac{2x^3z}{5y^3}\right)^2$

$= \dfrac{(-5)^3x^9y^6}{z^4} \times \dfrac{2^2x^6z^2}{5^2y^6}$

$= \dfrac{-5x^9}{z^2} \times \dfrac{4x^6}{1}$

$= \dfrac{-20x^{15}}{z^2}$

3 **a** $\left(\dfrac{-9m^3n}{2mp^3}\right)^2 \div \left(\dfrac{-3mn^3}{p^2}\right)^2$

$= \dfrac{(-9)^2m^6n^2}{2^2m^2p^6} \div \dfrac{(-3)^2m^2n^6}{p^4}$

$= \dfrac{81m^4n^2}{4p^6} \div \dfrac{9m^2n^6}{p^4}$

$= \dfrac{81m^4n^2}{4p^6} \times \dfrac{p^4}{9m^2n^6}$

$= \dfrac{9m^2}{4p^2} \times \dfrac{1}{n^4}$

$= \dfrac{9m^2}{4p^2n^4}$

b $\dfrac{6}{7}\left(\dfrac{7r^4st^4}{2s^4t^3}\right)^2 \div \left(\dfrac{-6rs}{r^2}\right)^2$

$= \dfrac{6}{7}\left(\dfrac{7^2r^8s^2t^8}{2^2s^8t^6}\right) \div \dfrac{(-6)^2r^2s^2}{r^4}$

$= \dfrac{6}{7}\left(\dfrac{49r^8t^2}{4s^6}\right) \div \dfrac{36s^2}{r^2}$

$= \dfrac{6}{7}\left(\dfrac{49r^8t^2}{4s^6}\right) \times \dfrac{r^2}{36s^2}$

$= \dfrac{7r^8t^2}{4s^6} \times \dfrac{r^2}{6s^2}$

$= \dfrac{7r^{10}t^2}{24s^8}$

Negative Powers

Multiple Choice

1 **Option B**

$$\frac{1}{8} = \frac{1}{2^3}$$

$$= 2^{-3}$$

2 **Option C**

$$\frac{1}{x^{-4}} = x^4$$

3 **Option C**

$$\left(2a^6b^{-4}\right)^{-3} = \frac{1}{(2a^6b^{-4})^3}$$

$$= \frac{1}{2^3a^{18}b^{-12}}$$

$$= \frac{b^{12}}{8a^{18}}$$

4 **Option B**

$$\left(\frac{3^4}{2^3}\right)^{-2}$$

$$= \left(\frac{2^3}{3^4}\right)^2$$

$$= \frac{2^6}{3^8}$$

5 **Option A**

$$\left(\frac{x^2y}{y^{-3}}\right)^2 = \left(\frac{x^2yy^3}{1}\right)^2$$

$$= \left(x^2y^4\right)^2$$

$$= x^4y^8$$

Time To Practice

1 a 4^{-1}

$= \frac{1}{4}$

b 3^{-3}

$= \frac{1}{3^3}$

$= \frac{1}{27}$

c 5^{-3}

$= \frac{1}{5^3}$

$= \frac{1}{125}$

d $2^{-3} \times 2^2$

$= 2^{-3+2} = 2^{-1}$

$= \frac{1}{2}$

e $7^{-4} \times 7^2$

$= 7^{-4+2} = 7^{-2}$

$= \frac{1}{7^2}$

$= \frac{1}{49}$

f $\frac{3}{4^{-2}}$

$= 3 \times \frac{4^2}{1}$

$= 3 \times 16$

$= 48$

g $\frac{1}{a^{-3}}$

$= a^3$

h $b^{-4} \times b$

$= b^{-4+1} = b^{-3}$

$= \frac{1}{b^3}$

i $c^{-5} \div c^{-3}$

$= c^{-5-(-3)}$

$= c^{-5+3} = c^{-2}$

$= \frac{1}{c^2}$

j $\frac{1}{y^{-4} \times y}$

$= \frac{1}{y^{-4+1}} = \frac{1}{y^{-3}}$

$= y^3$

k $x^{-3} \times y \times x^2$

$= x^{-3+2}y$

$= x^{-1}y$

$= \frac{1}{x} \times y$

$= \frac{y}{x}$

l $x^{-4} \times y^{-3} \times z^{-2}$

$= \frac{1}{x^4} \times \frac{1}{y^3} \times \frac{1}{z^2}$

$= \frac{1}{x^4y^3z^2}$

2 a $\left(\frac{5}{4}\right)^{-3}$

$=\left(\frac{4}{5}\right)^{3}$

$=\frac{4^3}{5^3}$

$=\frac{64}{125}$

b $\left(\frac{2}{3}\right)^{-2}$

$=\left(\frac{3}{2}\right)^{2}$

$=\frac{3^2}{2^2}$

$=\frac{9}{4}$

c $\left(\frac{1}{2^{-3}}\right)^{-2}$

$=\left(2^{-3}\right)^{2}$

$=2^{-6}$

$=\frac{1}{2^6}$

$=\frac{1}{64}$

d $\left(\frac{x^2}{y^{-3}}\right)^{4}$

$=\left(\frac{x^2y^3}{1}\right)^{4}$

$=x^8y^{12}$

e $\left(\frac{2x^{-3}}{y^{-2}}\right)^{-1}$

$=\left(\frac{2y^2}{x^3}\right)^{-1}$

$=\left(\frac{x^3}{2y^2}\right)^{1}$

$=\frac{x^3}{2y^2}$

f $\left(\frac{2}{x}\right)^{-3}\times x^{-2}$

$=\left(\frac{x}{2}\right)^{3}\times\frac{1}{x^2}$

$=\frac{x^3}{8}\times\frac{1}{x^2}$

$=\frac{x}{8}$

g $\frac{p^{-3}q^2}{r^{-4}p^2}$

$=\frac{q^2r^4}{p^3p^2}$

$=\frac{q^2r^4}{p^5}$

h $\frac{(p^{-1})^3\times\left(q^3\right)^{-3}}{p^{-6}\times(q^{-5})^3}$

$=\frac{p^{-3}\times q^{-9}}{p^{-6}\times q^{-15}}$

$=\frac{p^6\times q^{15}}{p^3\times q^9}$

$=p^3q^6$

i $\frac{\left(p^2\right)^{-2}\times\left(q^{-3}\right)^{-2}}{p^4\times(q^{-1})^2}$

$=\frac{p^{-4}\times q^6}{p^4\times q^{-2}}$

$=\frac{q^6\times q^2}{p^4\times p^4}$

$=\frac{q^8}{p^8}$

2 **j** $$\frac{x^{-3}y^{-4}}{z^{-2}} \times \frac{x^{-2}y^{3}}{z^{-2}}$$

$$= \frac{z^{2}}{x^{3}y^{4}} \times \frac{y^{3}z^{2}}{x^{2}}$$

$$= \frac{z^{4}}{x^{5}y}$$

k $$\left(\frac{x^{-2}y}{z^{2}}\right)^{3} \times \left(\frac{z}{x^{-1}}\right)^{-3}$$

$$= \left(\frac{y}{x^{2}z^{2}}\right)^{3} \times \left(\frac{xz}{1}\right)^{-3}$$

$$= \left(\frac{y}{x^{2}z^{2}}\right)^{3} \times \left(\frac{1}{xz}\right)^{3}$$

$$= \frac{y^{3}}{x^{6}z^{6}} \times \frac{1}{x^{3}z^{3}}$$

$$= \frac{y^{3}}{x^{9}z^{9}}$$

l $$\left(\frac{rs}{r^{-5}}\right)^{-3} \div \left(\frac{2r^{-8}}{s^{2}}\right)^{-2}$$

$$= \left(\frac{rsr^{5}}{1}\right)^{-3} \div \left(\frac{2}{r^{8}s^{2}}\right)^{-2}$$

$$= \left(\frac{1}{r^{6}s}\right)^{3} \div \left(\frac{r^{8}s^{2}}{2}\right)^{2}$$

$$= \frac{1}{r^{18}s^{3}} \div \frac{r^{16}s^{4}}{4}$$

$$= \frac{1}{r^{18}s^{3}} \times \frac{4}{r^{16}s^{4}}$$

$$= \frac{4}{r^{34}s^{7}}$$

3 **a** $\left(\frac{x^{-3}}{2^{-2}y^{-4}}\right)^{-2}$

$= \left(\frac{2^2y^4}{x^3}\right)^{-2}$

$= \left(\frac{x^3}{4y^4}\right)^{2}$

$= \frac{x^6}{4^2y^8}$

$= \frac{x^6}{16y^8}$

b $\left(\frac{m^{-2}n^{-3}}{3p^2}\right)^{-2} \div p^{-6}$

$= \left(\frac{1}{3p^2m^2n^3}\right)^{-2} \div \frac{1}{p^6}$

$= \left(\frac{3p^2m^2n^3}{1}\right)^{2} \div \frac{1}{p^6}$

$= \frac{9p^4m^4n^6}{1} \div \frac{1}{p^6}$

$= \frac{9p^4m^4n^6}{1} \times \frac{p^6}{1}$

$= 9p^{10}m^4n^6$

c $\left(\frac{2^{-2}x^{-1}}{y^3}\right)^{-2} \div y^{-9}$

$= \left(\frac{1}{2^2xy^3}\right)^{-2} \div \frac{1}{y^9}$

$= \left(\frac{2^2xy^3}{1}\right)^{2} \div \frac{1}{y^9}$

$= \frac{2^4x^2y^6}{1} \div \frac{1}{y^9}$

$= \frac{16x^2y^6}{1} \times \frac{y^9}{1}$

$= 16x^2y^{15}$

Challenge Yourself

1 **a**

$$\frac{2^{-4}x^{-3}y^2}{z^{-2}} \times \frac{x^6y^4}{z^{-4}}$$

$$= \frac{y^2z^2}{2^4x^3} \times \frac{x^6y^4z^4}{1}$$

$$= \frac{x^6y^6z^6}{16x^3}$$

$$= \frac{x^3y^6z^6}{16}$$

b

$$\frac{2^{-3}x^2y^{-4}}{z^{-3}} \times \left(\frac{x^{-1}z}{y^{-2}}\right)^{-2}$$

$$= \frac{x^2z^3}{2^3y^4} \times \left(\frac{y^2z}{x}\right)^{-2}$$

$$= \frac{x^2z^3}{8y^4} \times \left(\frac{x}{y^2z}\right)^2$$

$$= \frac{x^2z^3}{8y^4} \times \frac{x^2}{y^4z^2}$$

$$= \frac{x^4z}{8y^8}$$

2 **a**

$$\frac{(m^{-4})^3 \times (n^{-3})^2}{(2m^2n^3)^2} \times \left(\frac{mn}{3m^{-3}}\right)^{-2}$$

$$= \frac{m^{-12} \times n^{-6}}{2^2m^4n^6} \times \left(\frac{mnm^3}{3}\right)^{-2}$$

$$= \frac{1}{2^2m^4n^6m^{12}n^6} \times \left(\frac{3}{m^4n}\right)^2$$

$$= \frac{1}{4m^{16}n^{12}} \times \frac{9}{m^8n^2}$$

$$= \frac{9}{4m^{24}n^{14}}$$

b

$$\frac{4}{7}\left(\frac{m^3n^{-4}}{p^{-2}}\right)^{-2} \times \left(\frac{p^2m^{-1}}{n^3}\right)^{-3}$$

$$= \frac{4}{7}\left(\frac{m^3p^2}{n^4}\right)^{-2} \times \left(\frac{p^2}{n^3m}\right)^{-3}$$

$$= \frac{4}{7}\left(\frac{n^4}{m^3p^2}\right)^2 \times \left(\frac{n^3m}{p^2}\right)^3$$

$$= \frac{4}{7}\left(\frac{n^8}{m^6p^4}\right) \times \frac{n^9m^3}{p^6}$$

$$= \frac{4n^8}{7m^3p^4} \times \frac{n^9}{p^6}$$

$$= \frac{4n^{17}}{7m^3p^{10}}$$

3 **a** $\frac{2}{5}\left(\frac{3x^{-3}y^{-1}}{x^{-2}y^{-4}z}\right)^{-2} \div \left(\frac{x^2y^{-1}}{4y^5z^3}\right)^{-1}$

$= \frac{2}{5}\left(\frac{3x^2y^4}{x^3yz}\right)^{-2} \div \left(\frac{x^2}{4y^5zy}\right)^{-1}$

$= \frac{2}{5}\left(\frac{3x^2y^4}{x^3yz}\right)^{-2} \div \left(\frac{x^2}{4y^6z}\right)^{-1}$

$= \frac{2}{5}\left(\frac{3y^3}{xz}\right)^{-2} \div \left(\frac{x^2}{4y^6z}\right)^{-1}$

$= \frac{2}{5}\left(\frac{xz}{3y^3}\right)^{2} \div \left(\frac{4y^6z}{x^2}\right)^{1}$

$= \frac{2}{5}\left(\frac{x^2z^2}{9y^6}\right) \div \frac{4y^6z}{x^2}$

$= \frac{2x^2z^2}{45y^6} \times \frac{x^2}{4y^6z}$

$= \frac{x^2}{45y^6} \times \frac{x^2}{2y^6z}$

$= \frac{x^4}{90y^{12}z}$

b $\left(\frac{x^{-2}y^{-1}}{x^{-4}y^{-1}z^{-5}}\right)^{-2} \div \left(\frac{x^{-2}y^{-5}z^{-9}}{3^{-1}y^{-8}z^{-1}}\right)^{-1}$

$= \left(\frac{x^4yz^5}{x^2y}\right)^{-2} \div \left(\frac{3y^8z}{x^2y^5z^9}\right)^{-1}$

$= \left(\frac{x^4z^5}{x^2}\right)^{-2} \div \left(\frac{3y^3}{x^2z^8}\right)^{-1}$

$= \left(\frac{x^2}{x^4z^5}\right)^{2} \div \left(\frac{x^2z^8}{3y^3}\right)^{1}$

$= \frac{x^4}{x^8z^{10}} \times \frac{3y^3}{x^2z^8}$

$= \frac{1}{x^4z^{10}} \times \frac{3y^3}{x^2z^8}$

$= \frac{3y^3}{x^6z^{18}}$

Fractional Powers

Multiple Choice

1 **Option A**

$16^{\frac{1}{4}}$

$= (2^4)^{\frac{1}{4}}$

$= 2^1$

$= 2$

2 **Option C**

$32^{\frac{3}{5}}$

$= \left(32^{\frac{1}{5}}\right)^3$

$= (2)^3$

$= 8$

3 **Option B**

$\left(\frac{64}{x^3}\right)^{\frac{1}{3}}$

$= \frac{64^{\frac{1}{3}}}{x^1}$

$= \frac{4}{x}$

4 **Option D**

$(81y^{16})^{\frac{1}{4}}$

$= 81^{\frac{1}{4}}y^4$

$= 3y^4$

5 **Option B**

$(a^9b^6)^{\frac{2}{3}}$

$= a^{\frac{18}{3}}b^{\frac{12}{3}}$

$= a^6b^4$

Time To Practice

1 a $27^{\frac{1}{3}}$

$= 3$

b $121^{\frac{1}{2}}$

$= 11$

c $49^{\frac{1}{2}}$

$= 7$

d $8^{\frac{5}{3}}$

$= \left(8^{\frac{1}{3}}\right)^5 = (2)^5 = 32$

e $25^{\frac{3}{2}}$

$= \left(25^{\frac{1}{2}}\right)^3$

$= (5)^3 = 125$

f $49^{\frac{3}{2}}$

$= \left(49^{\frac{1}{2}}\right)^3$

$= (7)^3 = 343$

g $16^{\frac{3}{4}}$

$= \left(16^{\frac{1}{4}}\right)^3$

$= (2)^3 = 8$

h $4^{-\frac{1}{2}}$

$= \frac{1}{4^{\frac{1}{2}}}$

$= \frac{1}{2}$

i $27^{-\frac{2}{3}}$

$= \frac{1}{27^{\frac{2}{3}}} = \frac{1}{\left(27^{\frac{1}{3}}\right)^2}$

$= \frac{1}{(3)^2} = \frac{1}{9}$

j $8^{-\frac{1}{3}}$

$= \frac{1}{8^{\frac{1}{3}}} = \frac{1}{2}$

k $64^{-\frac{2}{3}}$

$= \frac{1}{64^{\frac{2}{3}}} = \frac{1}{\left(64^{\frac{1}{3}}\right)^2}$

$= \frac{1}{(4)^2} = \frac{1}{16}$

l $(125^{-1})^{\frac{1}{3}}$

$= \left(\frac{1}{125}\right)^{\frac{1}{3}}$

$= \frac{1}{125^{\frac{1}{3}}} = \frac{1}{5}$

2 **a** $\left(\frac{9}{16}\right)^{-\frac{1}{2}}$

$= \left(\frac{16}{9}\right)^{\frac{1}{2}}$

$= \frac{16^{\frac{1}{2}}}{9^{\frac{1}{2}}} = \frac{4}{3}$

b $\left(\frac{16}{81}\right)^{-\frac{3}{4}}$

$= \left(\frac{81}{16}\right)^{\frac{3}{4}} = \frac{81^{\frac{3}{4}}}{16^{\frac{3}{4}}}$

$= \frac{\left(81^{\frac{1}{4}}\right)^3}{\left(16^{\frac{1}{4}}\right)^3}$

$= \frac{3^3}{2^3} = \frac{27}{8}$

c $\left(\frac{27}{64}\right)^{-\frac{2}{3}}$

$= \left(\frac{64}{27}\right)^{\frac{2}{3}} = \frac{64^{\frac{2}{3}}}{27^{\frac{2}{3}}}$

$= \frac{\left(64^{\frac{1}{3}}\right)^2}{\left(27^{\frac{1}{3}}\right)^2} = \frac{4^2}{3^2} = \frac{16}{9}$

3 **a** $\sqrt{x^3}$

$= x^{\frac{2}{3}}$

b $\sqrt[3]{x^5}$

$= x^{\frac{5}{3}}$

c $\sqrt[4]{16x}$

$= (16x)^{\frac{1}{4}}$

d $\sqrt[3]{y^2}$

$= \left(y^2\right)^{\frac{1}{3}} = y^{\frac{2}{3}}$

e $\sqrt[6]{64y^7}$

$= (64y^7)^{\frac{1}{6}}$

f $\sqrt[3]{8y^4}$

$= (8y^4)^{\frac{1}{3}}$

4 **a** $\left(x^6\right)^{\frac{1}{3}}$

$= x^{\frac{6}{3}} = x^2$

b $\left(25x^2\right)^{\frac{1}{2}}$

$= 25^{\frac{1}{2}}x = 5x$

c $\left(16x^8\right)^{\frac{1}{4}}$

$16^{\frac{1}{4}}x^2 = 2x^2$

d $\left(a^2b^4\right)^{\frac{1}{2}}$

$= ab^2$

e $\left(8a^3b^3\right)^{\frac{1}{3}}$

$= 8^{\frac{1}{3}}ab$

$= 2ab$

f $3a^{\frac{1}{2}} \times 2a^{-\frac{3}{4}}$

$= 6a^{\frac{1}{2}+(-\frac{3}{4})}$

$= 6a^{\frac{1}{2}+(-\frac{3}{4})}$

$= 6a^{-\frac{1}{4}}$

$= \frac{6}{a^{\frac{1}{4}}}$

g $\left(a^8\right)^{-\frac{1}{2}}$

$= a^{-4}$

$= \frac{1}{a^4}$

h $\left(8x^9\right)^{-\frac{1}{3}}$

$= 8^{-\frac{1}{3}}a^{-3}$

$= \frac{1}{8^{\frac{1}{3}}a^3}$

$= \frac{1}{2a^3}$

i $\left(x^4y^8\right)^{-\frac{3}{4}}$

$= x^{-3}y^{-6}$

$= \frac{1}{x^3y^6}$

j $\left(x^2\right)^{\frac{1}{2}} \times x^{-\frac{5}{3}}$

$= x \times x^{-\frac{5}{3}}$

$= x^{1+(-\frac{5}{3})}$

$= x^{-\frac{2}{3}}$

$= \frac{1}{x^{\frac{2}{3}}}$

k $\left(\frac{x^{12}y^{-8}}{z^4}\right)^{\frac{3}{4}}$

$= \left(\frac{x^{12}}{y^8z^4}\right)^{\frac{3}{4}}$

$= \frac{x^9}{y^6z^3}$

l $\left(\frac{x^6}{y^{-9}}\right)^{\frac{2}{3}}$

$= \left(x^6y^9\right)^{\frac{2}{3}}$

$= x^4y^6$

Challenge Yourself

1 **a**

$$\left(a^2b^4\right)^{-\frac{1}{2}} \times \left(ba^2\right)^{\frac{3}{2}}$$

$$= a^{-1}b^{-2} \times b^{\frac{3}{2}}a^3$$

$$= a^{-1+3}b^{-2+\frac{3}{2}}$$

$$= a^2b^{-\frac{1}{2}}$$

$$= \frac{a^2}{b^{\frac{1}{2}}}$$

b

$$\left(a^3b^{-2}\right)^{-\frac{2}{3}} \times \left(a^{-1}b^2\right)^{\frac{1}{3}}$$

$$= a^{-2}b^{\frac{4}{3}} \times a^{-\frac{1}{3}}b^{\frac{2}{3}}$$

$$= a^{-2-\frac{1}{3}}b^{\frac{4}{3}+\frac{2}{3}}$$

$$= a^{-\frac{7}{3}}b^2$$

$$= \frac{b^2}{a^{\frac{7}{3}}}$$

2 **a**

$$\left(2^{-1}x^{\frac{3}{2}}y^{\frac{5}{4}}\right)^2 \times \left(8x^{-\frac{7}{4}}y^{\frac{1}{2}}\right)^{\frac{2}{3}}$$

$$= 2^{-2}x^3y^{\frac{5}{2}} \times 8^{\frac{2}{3}}x^{-\frac{7}{6}}y^{\frac{1}{3}}$$

$$= \frac{1}{2^2} \times \left(8^{\frac{1}{3}}\right)^2 \times x^{3-\frac{7}{6}}y^{\frac{5}{2}+\frac{1}{3}}$$

$$= \frac{1}{4} \times (2)^2 \times x^{\frac{11}{6}}y^{\frac{17}{6}}$$

$$= x^{\frac{11}{6}}y^{\frac{17}{6}}$$

b

$$\left(x^{\frac{2}{5}}y^{-\frac{1}{4}}\right)^2 \div \left(x^{-\frac{3}{2}}y^{\frac{1}{3}}\right)^{\frac{2}{3}}$$

$$= x^{\frac{4}{5}}y^{-\frac{1}{2}} \div x^{-1}y^{\frac{2}{9}}$$

$$= x^{\frac{4}{5}-(-1)}y^{-\frac{1}{2}-\frac{2}{9}}$$

$$= x^{\frac{9}{5}}y^{-\frac{13}{18}}$$

$$= \frac{x^{\frac{9}{5}}}{y^{\frac{13}{18}}}$$

3 **a** $\left(\dfrac{a^{\frac{1}{2}}b^{\frac{1}{3}}}{2c^{-\frac{1}{2}}}\right)^2 \times \left(a^{-1}c^{\frac{1}{3}}\right)^{-1}$

$= \dfrac{ab^{\frac{2}{3}}}{4c^{-1}} \times ac^{-\frac{1}{3}}$

$= \dfrac{ab^{\frac{2}{3}}c}{4} \times ac^{-\frac{1}{3}}$

$= \dfrac{a^2b^{\frac{2}{3}}c^{1-\frac{1}{3}}}{4}$

$= \dfrac{a^2b^{\frac{2}{3}}c^{\frac{2}{3}}}{4}$

b $\left(\dfrac{2a^2b^{-1}}{c^{\frac{1}{2}}}\right)^{-2} \div \left(\dfrac{4a^{-1}b^2}{c}\right)^{\frac{1}{2}}$

$= \left(\dfrac{2a^2}{bc^{\frac{1}{2}}}\right)^{-2} \div \left(\dfrac{4b^2}{ac}\right)^{\frac{1}{2}}$

$= \left(\dfrac{bc^{\frac{1}{2}}}{2a^2}\right)^2 \div \dfrac{2b}{a^{\frac{1}{2}}c^{\frac{1}{2}}}$

$= \dfrac{b^2c}{4a^4} \times \dfrac{a^{\frac{1}{2}}c^{\frac{1}{2}}}{2b}$

$= \dfrac{bc^{\frac{3}{2}}}{8a^{\frac{7}{2}}}$

Surds

Identifying Surds

Multiple Choice

1 **Option C**

No number multiplied by itself equals 50, $\therefore \sqrt{50}$ is a surd

2 **Option B**

$\sqrt[3]{64} = 4$

$\therefore$ Not a surd. Option B is NOT true.

3 **Option D**

$\sqrt[3]{27} = 3$

$\sqrt[3]{1000} = 10$

$\sqrt[3]{125} = 5$

$\therefore \sqrt[3]{20}$ is a surd.

4 **Option A**

$$\sqrt{0.001} = \sqrt{\frac{1}{1000}}$$

$$= \sqrt{\frac{1}{100 \times 10}}$$

$$= \frac{1}{10}\sqrt{\frac{1}{10}}$$

$\therefore \sqrt{0.001}$ is a surd.

5 **Option D**

$$\sqrt{\frac{5}{25}} = \sqrt{\frac{1}{5}}$$

No number multiplied by itself equals 5, $\therefore \sqrt{\frac{5}{25}}$ is a surd.

Time To Practice

1 a $\sqrt{10}$

No number multiplied by itself equals 10.

$\therefore$ Yes, it's a surd.

b $\sqrt{144} = \sqrt{12 \times 12} = 12$

$\therefore$ NOT a surd.

c $\sqrt{55}$

No number multiplied by itself equals 55.

$\therefore$ Yes, it's a surd.

d $\sqrt{18}$

No number multiplied by itself equals 18.

$\therefore$ Yes, it's a surd.

e $\sqrt{8}$

No number multiplied by itself equals 8.

$\therefore$ Yes, it's a surd.

f $\sqrt{1} = \sqrt{1 \times 1} = 1$

$\therefore$ NOT a surd.

g $\sqrt[3]{21}$

No number multiplied by itself three times equals 21.

$\therefore$ Yes, it's a surd.

h $\sqrt[3]{216} = \sqrt[3]{6 \times 6 \times 6} = 6$

$\therefore$ NOT a surd.

i $\sqrt[3]{32}$

No number multiplied by itself three times equals 32.

$\therefore$ Yes, it's a surd.

j $\sqrt[4]{16} = \sqrt[4]{2 \times 2 \times 2 \times 2} = 2$

$\therefore$ NOT a surd.

1 k $\sqrt[4]{40}$

No number multiplied by itself four times equals 40.

$\therefore$ Yes, it's a surd.

l $\sqrt[5]{32} = \sqrt[5]{2 \times 2 \times 2 \times 2 \times 2} = 2$

$\therefore$ NOT a surd.

m $\sqrt{\frac{1}{4}} = \sqrt{\frac{1}{2} \times \frac{1}{2}} = \frac{1}{2}$

$\therefore$ NOT a surd.

n $\sqrt{\frac{4}{49}} = \sqrt{\frac{2}{7} \times \frac{2}{7}} = \frac{2}{7}$

$\therefore$ NOT a surd.

o $\sqrt{\frac{16}{32}}$

No number multiplied by itself equals 32.

$\therefore$ Yes, it's a surd.

p $\sqrt{0.36} = \sqrt{\frac{36}{100}} = \sqrt{\frac{6}{10} \times \frac{6}{10}}$

$= \frac{6}{10} = \frac{3}{5}$

$\therefore$ NOT a surd.

q $\sqrt{0.5} = \sqrt{\frac{50}{100}}$

No number multiplied by itself two times equals 50.

$\therefore$ Yes, it's a surd.

r $\sqrt{0.65} = \sqrt{\frac{65}{100}} = \sqrt{\frac{17}{20}}$

No number multiplied by itself two times equals 17 or 20.

$\therefore$ Yes, it's a surd.

Challenge Yourself

1 **a** $\sqrt[3]{\frac{1}{8}}$

$= \sqrt[3]{\frac{1}{2} \times \frac{1}{2} \times \frac{1}{2}} = \frac{1}{2}$

∴ NOT a surd.

b $\sqrt[4]{\frac{1}{32}}$

No number multiplied by itself four times equals 32.

∴ Yes, it's a surd.

c $\sqrt[3]{\frac{64}{125}}$

$= \sqrt[3]{\frac{4}{5} \times \frac{4}{5} \times \frac{4}{5}} = \frac{4}{5}$

∴ NOT a surd.

d $\sqrt[3]{0.125}$

$= \sqrt[3]{\frac{125}{1000}} = \sqrt[3]{\frac{5}{10} \times \frac{5}{10} \times \frac{5}{10}} = \frac{5}{10}$

∴ NOT a surd.

e $\sqrt[3]{0.1} == \sqrt[3]{\frac{1}{10}}$

No number multiplied by itself three times equals 10.

∴ Yes, it's a surd.

f $\sqrt[3]{0.008}$

$= \sqrt[3]{\frac{8}{1000}} = \sqrt[3]{\frac{2}{10} \times \frac{2}{10} \times \frac{2}{10}} = \frac{2}{10}$

∴ NOT a surd.

2

$\sqrt{\frac{1}{8}}$ → No number multiplied by itself equals 8.

$\sqrt{21}$ → No number multiplied by itself equals 21.

Simplifying Surds

Multiple Choice

1 **Option A**

$\sqrt{50}$

$= \sqrt{25 \times 2}$

$= \sqrt{25} \times \sqrt{2}$

$= 5\sqrt{2}$

2 **Option C**

$\sqrt{200}$

$= \sqrt{100 \times 2}$

$= \sqrt{100} \times \sqrt{2}$

$= 10\sqrt{2}$

3 **Option D**

$3\sqrt{27}$

$= 3\sqrt{9 \times 3}$

$= 3\sqrt{9} \times \sqrt{3}$

$= 3 \times 3 \times \sqrt{3}$

$= 9\sqrt{3}$

4 **Option B**

$\frac{3}{4}\sqrt{80}$

$= \frac{3}{4}\sqrt{16 \times 5}$

$= \frac{3}{4}\sqrt{16} \times \sqrt{5}$

$= \frac{3}{4} \times 4 \times \sqrt{5}$

$= 3\sqrt{5}$

5 **Option A**

$6\sqrt{3}$

$= \sqrt{36 \times 3}$

$= \sqrt{108}$

$\therefore 108$

Time To Practice

1 a $\sqrt{12}$

$= \sqrt{4 \times 3} = \sqrt{4} \times \sqrt{3}$

$= 2\sqrt{3}$

b $\sqrt{18}$

$= \sqrt{9 \times 2} = \sqrt{9} \times \sqrt{2}$

$= 3\sqrt{2}$

c $\sqrt{20}$

$= \sqrt{4 \times 5} = \sqrt{4} \times \sqrt{5}$

$= 2\sqrt{5}$

d $\sqrt{24}$

$= \sqrt{4 \times 6} = \sqrt{4} \times \sqrt{6}$

$= 2\sqrt{6}$

e $\sqrt{27}$

$= \sqrt{9 \times 3} = \sqrt{9} \times \sqrt{3}$

$= 3\sqrt{3}$

f $\sqrt{72}$

$= \sqrt{36 \times 2} = \sqrt{36} \times \sqrt{2}$

$= 6\sqrt{2}$

g $\sqrt{98}$

$= \sqrt{49 \times 2} = \sqrt{49} \times \sqrt{2}$

$= 7\sqrt{2}$

h $\sqrt{125}$

$= \sqrt{25 \times 5} = \sqrt{25} \times \sqrt{5}$

$= 5\sqrt{5}$

i $\sqrt{360}$

$= \sqrt{36 \times 10} = \sqrt{36} \times \sqrt{10}$

$= 6\sqrt{10}$

2 **a** $2\sqrt{18}$

$= 2\sqrt{9 \times 2} = 2\sqrt{9} \times \sqrt{2}$

$= 2 \times 3 \times \sqrt{2}$

$= 6\sqrt{2}$

b $3\sqrt{20}$

$= 3\sqrt{4 \times 5} = 3\sqrt{4} \times \sqrt{5}$

$= 3 \times 2 \times \sqrt{5}$

$= 6\sqrt{5}$

c $2\sqrt{63}$

$= 2\sqrt{9 \times 7} = 2\sqrt{9} \times \sqrt{7}$

$= 2 \times 3 \times \sqrt{7}$

$= 6\sqrt{7}$

d $4\sqrt{48}$

$= 4\sqrt{16 \times 3} = 4\sqrt{16} \times \sqrt{3}$

$= 4 \times 4 \times \sqrt{3}$

$= 16\sqrt{3}$

e $5\sqrt{98}$

$= 5\sqrt{49 \times 2} = 5\sqrt{49} \times \sqrt{2}$

$= 5 \times 7 \times \sqrt{2}$

$= 35\sqrt{2}$

f $2\sqrt{800}$

$= 2\sqrt{400 \times 2} = 2\sqrt{400} \times \sqrt{2}$

$= 2 \times 20 \times \sqrt{2}$

$= 40\sqrt{2}$

g $4\sqrt{150}$

$= 4\sqrt{25 \times 6} = 4\sqrt{25} \times \sqrt{6}$

$= 4 \times 5 \times \sqrt{6}$

$= 20\sqrt{6}$

h $3\sqrt{147}$

$= 3\sqrt{49 \times 3} = 3\sqrt{49} \times \sqrt{3}$

$= 3 \times 7 \times \sqrt{3}$

$= 21\sqrt{3}$

i $6\sqrt{75}$

$= 6\sqrt{25 \times 3} = 6\sqrt{25} \times \sqrt{3}$

$= 6 \times 5 \times \sqrt{3}$

$= 30\sqrt{3}$

3 **a** $\frac{1}{2}\sqrt{20}$

$= \frac{1}{2}\sqrt{4 \times 5}$

$= \frac{1}{2}\sqrt{4} \times \sqrt{5}$

$= \frac{1}{2} \times 2 \times \sqrt{5}$

$= \sqrt{5}$

b $\frac{1}{3}\sqrt{28}$

$= \frac{1}{3}\sqrt{4 \times 7}$

$= \frac{1}{3}\sqrt{4} \times \sqrt{7}$

$= \frac{1}{3} \times 2 \times \sqrt{7}$

$= \frac{2\sqrt{7}}{3}$

c $\frac{1}{4}\sqrt{54}$

$= \frac{1}{4}\sqrt{9 \times 6}$

$= \frac{1}{4}\sqrt{9} \times \sqrt{6}$

$= \frac{1}{4} \times 3 \times \sqrt{6}$

$= \frac{3\sqrt{6}}{4}$

d $\frac{1}{6}\sqrt{24}$

$= \frac{1}{6}\sqrt{4 \times 6}$

$= \frac{1}{6}\sqrt{4} \times \sqrt{6}$

$= \frac{1}{6} \times 2 \times \sqrt{6}$

$= \frac{2\sqrt{6}}{6} = \frac{\sqrt{6}}{3}$

e $\frac{1}{5}\sqrt{80}$

$= \frac{1}{5}\sqrt{16 \times 5}$

$= \frac{1}{5}\sqrt{16} \times \sqrt{5}$

$= \frac{1}{5} \times 4 \times \sqrt{5}$

$= \frac{4\sqrt{5}}{5}$

f $\frac{1}{9}\sqrt{99}$

$= \frac{1}{9}\sqrt{9 \times 11}$

$= \frac{1}{9}\sqrt{9} \times \sqrt{11}$

$= \frac{1}{9} \times 3 \times \sqrt{11}$

$= \frac{3\sqrt{11}}{9} = \frac{\sqrt{11}}{3}$

3 **g** $\frac{1}{8}\sqrt{84}$

$= \frac{1}{8}\sqrt{4 \times 21} = \frac{1}{8}\sqrt{4} \times \sqrt{21}$

$= \frac{1}{8} \times 2 \times \sqrt{21}$

$= \frac{2\sqrt{21}}{8}$

$= \frac{\sqrt{21}}{4}$

h $\frac{1}{7}\sqrt{56}$

$= \frac{1}{7}\sqrt{4 \times 14} = \frac{1}{7}\sqrt{4} \times \sqrt{14}$

$= \frac{1}{7} \times 2 \times \sqrt{14}$

$= \frac{2\sqrt{14}}{7}$

i $\frac{1}{3}\sqrt{68}$

$= \frac{1}{3}\sqrt{4 \times 17} = \frac{1}{3}\sqrt{4} \times \sqrt{17}$

$= \frac{1}{3} \times 2 \times \sqrt{17}$

$= \frac{2\sqrt{17}}{3}$

4 **a** $5\sqrt{2}$

$= \sqrt{5^2 \times 2} = \sqrt{25 \times 2}$

$= \sqrt{50}$

b $3\sqrt{3}$

$= \sqrt{3^2 \times 3} = \sqrt{9 \times 3}$

$= \sqrt{27}$

c $8\sqrt{2}$

$= \sqrt{8^2 \times 2} = \sqrt{64 \times 2}$

$= \sqrt{128}$

d $10\sqrt{7}$

$= \sqrt{10^2 \times 7} = \sqrt{100 \times 7}$

$= \sqrt{700}$

4 **e** $9\sqrt{10}$

$= \sqrt{9^2 \times 10} = \sqrt{81 \times 10}$

$= \sqrt{810}$

f $3\sqrt{11}$

$= \sqrt{3^2 \times 11} = \sqrt{9 \times 11}$

$= \sqrt{99}$

Challenge Yourself

1 $\sqrt{432} = 12\sqrt{k}$

$\sqrt{432} = \sqrt{144 \times 3}$

$= \sqrt{144} \times \sqrt{3}$

$= 12\sqrt{3}$

$\therefore k = 3$

2 $\sqrt{450} = c\sqrt{2}$

$\sqrt{450} = \sqrt{225 \times 2}$

$= \sqrt{225} \times \sqrt{2}$

$= 15\sqrt{2}$

$\therefore c = 15$

3 $\sqrt{1200} = a\sqrt{b}$

$= \sqrt{400 \times 3} = \sqrt{400} \times \sqrt{3}$

$= 20\sqrt{3}$

$\therefore a = 20 \text{ and } b = 3$

4 **a** $\frac{2}{7}\sqrt{98}$

$= \frac{2}{7}\sqrt{49 \times 2}$

$= \frac{2}{7}\sqrt{49} \times \sqrt{2}$

$= \frac{2}{7} \times 7 \times \sqrt{2}$

$= 2\sqrt{2}$

b $\frac{1}{9}\sqrt{162}$

$= \frac{1}{9}\sqrt{81 \times 2}$

$= \frac{1}{9}\sqrt{81} \times \sqrt{2}$

$= \frac{1}{9} \times 9 \times \sqrt{2}$

$= \sqrt{2}$

c $\frac{3}{10}\sqrt{175}$

$= \frac{3}{10}\sqrt{25 \times 7}$

$= \frac{3}{10}\sqrt{25} \times \sqrt{7}$

$= \frac{3}{10} \times 5 \times \sqrt{7}$

$= \frac{15\sqrt{7}}{10}$

$= \frac{3\sqrt{7}}{2}$

d $\frac{3}{14}\sqrt{147}$

$= \frac{3}{14}\sqrt{49 \times 3}$

$= \frac{3}{14}\sqrt{49} \times \sqrt{3}$

$= \frac{3}{14} \times 7 \times \sqrt{3}$

$= \frac{21\sqrt{7}}{14}$

$= \frac{3\sqrt{3}}{2}$

Square and Square Roots of Surds

Multiple Choice

1 **Option B**

$\left(\sqrt{5}\right)^2$

$= \sqrt{5} \times \sqrt{5}$

$= 5$

2 **Option D**

$\left(2\sqrt{4x}\right)^2$

$= 2^2 \times \left(\sqrt{4x}\right)^2$

$= 4 \times 4x$

$= 16x$

3 **Option C**

$\sqrt{81a^4}$

$= \sqrt{(9a^2)^2}$

$= 9a^2$

4 **Option C**

$\sqrt{100x^6y^4}$

$= \sqrt{(10x^3y^2)^2}$

$= 10x^3y^2$

5 **Option A**

$\sqrt{200x^5}$

$= \sqrt{100x^4 \times 2x}$

$= \sqrt{(10x^2)^2} \times \sqrt{2x}$

$= 10x^2\sqrt{2x}$

Time To Practice

1 **a** $\left(\sqrt{7}\right)^2$

$= \sqrt{7} \times \sqrt{7}$

$= 7$

b $\left(\sqrt{11}\right)^2$

$= \sqrt{11} \times \sqrt{11}$

$= 11$

c $\left(\sqrt{13}\right)^2$

$= \sqrt{13} \times \sqrt{13}$

$= 13$

d $\left(\sqrt{5x}\right)^2$

$= \sqrt{5x} \times \sqrt{5x}$

$= 5x$

e $\left(\sqrt{9x}\right)^2$

$= \sqrt{9x} \times \sqrt{9x}$

$= 9x$

f $\left(\sqrt{3x^2}\right)^2$

$= \sqrt{3x^2} \times \sqrt{3x^2}$

$= 3x^2$

g $\left(2\sqrt{2x}\right)^2$

$= 2^2 \times \left(\sqrt{2x}\right)^2$

$= 4 \times 2x$

$= 8x$

h $\left(-4\sqrt{3x}\right)^2$

$= (-4)^2 \times \left(\sqrt{3x}\right)^2$

$= 16 \times 3x$

$= 48x$

i $\left(-5\sqrt{2x^2}\right)^2$

$= (-5)^2 \times \left(\sqrt{2x^2}\right)^2$

$= 25 \times 2x^2$

$= 50x^2$

2 a $\sqrt{z^2}$

$= z$

b $\sqrt{25z^2}$

$= \sqrt{(5z)^2}$

$= 5z$

c $\sqrt{121z^2}$

$= \sqrt{(11z)^2}$

$= 11z$

d $\sqrt{36b^2}$

$= \sqrt{(6b)^2}$

$= 6b$

e $\sqrt{100b^2}$

$= \sqrt{(10b)^2}$

$= 10b$

f $\sqrt{49b^2}$

$= \sqrt{(7b)^2}$

$= 7b$

g $\sqrt{144p^2q^2}$

$= \sqrt{(12pq)^2}$

$= 12pq$

h $\sqrt{225p^6q^4}$

$= \sqrt{(15p^3q^2)^2}$

$= 15p^3q^2$

i $\sqrt{289p^2q^8}$

$= \sqrt{(17pq^4)^2}$

$= 17pq^4$

3 a $\sqrt{12x^2}$

$= \sqrt{4x^2 \times 3}$

$= \sqrt{(2x)^2 \times 3}$

$= \sqrt{(2x)^2} \times \sqrt{3}$

$= 2x\sqrt{3}$

b $\sqrt{28x^2}$

$= \sqrt{4x^2 \times 7}$

$= \sqrt{(2x)^2 \times 7}$

$= \sqrt{(2x)^2} \times \sqrt{7}$

$= 2x\sqrt{7}$

3 c $\sqrt{45y^2}$

$= \sqrt{9y^2 \times 5}$
$= \sqrt{(3y)^2 \times 5}$
$= \sqrt{(3y)^2} \times \sqrt{5}$
$= 3y\sqrt{5}$

d $\sqrt{48x^8}$

$= \sqrt{16x^8 \times 3}$
$= \sqrt{(4x^4)^2 \times 3}$
$= \sqrt{(4x^4)^2} \times \sqrt{3}$
$= 4x^4\sqrt{3}$

e $\sqrt{50x^6}$

$= \sqrt{25x^6 \times 2}$
$= \sqrt{(5x^3)^2 \times 2}$
$= \sqrt{(5x^3)^2} \times \sqrt{2}$
$= 5x^3\sqrt{2}$

f $\sqrt{72y^4}$

$= \sqrt{36y^4 \times 2}$
$= \sqrt{(6y^2)^2 \times 2}$
$= \sqrt{(6y^2)^2} \times \sqrt{2}$
$= 6y^2\sqrt{2}$

g $\sqrt{75x^8y^6}$

$= \sqrt{25x^8y^6 \times 3}$
$= \sqrt{(5x^4y^3)^2 \times 3}$
$= \sqrt{(5x^4y^3)^2} \times \sqrt{3}$
$= 5x^4y^3\sqrt{3}$

h $\sqrt{96x^4y^2}$

$= \sqrt{16x^4y^2 \times 6}$
$= \sqrt{(4x^2y)^2 \times 6}$
$= \sqrt{(4x^2y)^2} \times \sqrt{6}$
$= 4x^2y\sqrt{6}$

i $\sqrt{98x^2y^6}$

$= \sqrt{49x^2y^6 \times 2}$
$= \sqrt{(7xy^3)^2 \times 2}$
$= \sqrt{(7xy^3)^2} \times \sqrt{2}$
$= 7xy^3\sqrt{2}$

4 a $\sqrt{a^{13}}$

$= \sqrt{a^{12} \times a}$

$= \sqrt{(a^6)^2 \times a}$

$= \sqrt{(a^6)^2} \times \sqrt{a}$

$= a^6\sqrt{a}$

b $\sqrt{2b^5}$

$= \sqrt{b^4 \times 2b}$

$= \sqrt{(b^2)^2 \times 2b}$

$= \sqrt{(b^2)^2} \times \sqrt{2b}$

$= b^2\sqrt{2b}$

c $\sqrt{4a^7}$

$= \sqrt{4a^6 \times a}$

$= \sqrt{(2a^3)^2 \times a}$

$= \sqrt{(2a^3)^2} \times \sqrt{a}$

$= 2a^3\sqrt{a}$

d $3\sqrt{16a^9}$

$= 3\sqrt{16a^8 \times a}$

$= 3\sqrt{(4a^4)^2 \times a}$

$= 3\sqrt{(4a^4)^2} \times \sqrt{a}$

$= 3 \times 4a^4\sqrt{a}$

$= 12a^4\sqrt{a}$

e $2\sqrt{4a^3b}$

$= 2\sqrt{4a^2 \times ab}$

$= 2\sqrt{(2a)^2 \times ab}$

$= 2\sqrt{(2a)^2} \times \sqrt{ab}$

$= 2 \times 2a\sqrt{ab}$

$= 4a\sqrt{ab}$

f $\sqrt{45a^7b^5}$

$= \sqrt{9a^6b^4 \times 5ab}$

$= \sqrt{(3a^3b^2)^2 \times 5ab}$

$= \sqrt{(3a^3b^2)^2} \times \sqrt{5ab}$

$= 3a^3b^2\sqrt{5ab}$

4 **g** $3\sqrt{14ab^5}$

$= 3\sqrt{b^4 \times 14ab}$

$= 3\sqrt{(b^2)^2 \times 14ab}$

$= 3\sqrt{(b^2)^2} \times \sqrt{14ab}$

$= 3b^2\sqrt{14ab}$

h $4\sqrt{18a^{15}b^3}$

$= 4\sqrt{9a^{14}b^2 \times 2ab}$

$= 4\sqrt{(3a^7b)^2 \times 2ab}$

$= 4\sqrt{(3a^7b)^2} \times \sqrt{2ab}$

$= 4 \times 3a^7b \times \sqrt{2ab}$

$= 12a^7b\sqrt{2ab}$

i $5\sqrt{15a^5b^9}$

$= 5\sqrt{a^4b^8 \times 15ab}$

$= 5\sqrt{(a^2b^4)^2 \times 15ab}$

$= 5\sqrt{(a^2b^4)^2} \times \sqrt{15ab}$

$= 5 \times a^2b^4 \times \sqrt{15ab}$

$= 5a^2b^4\sqrt{15ab}$

Challenge Yourself

1 **a** $3\sqrt{125x^4y^4}$

$= 3\sqrt{25x^4y^4 \times 5}$

$= 3\sqrt{(5x^2y^2)^2 \times 5}$

$= 3\sqrt{(5x^2y^2)^2} \times \sqrt{5}$

$= 3 \times 5x^2y^2 \times \sqrt{5}$

$= 15x^2y^2\sqrt{5}$

b $7\sqrt{162x^2y^6}$

$= 7\sqrt{81x^2y^6 \times 2}$

$= 7\sqrt{(9xy^3)^2 \times 2}$

$= 7\sqrt{(9xy^3)^2} \times \sqrt{2}$

$= 7 \times 9xy^3 \times \sqrt{2}$

$= 63xy^3\sqrt{2}$

1 c $\frac{1}{2}\sqrt{88x^4y^{12}}$

$=\frac{1}{2}\sqrt{4x^4y^{12}\times 22}$

$=\frac{1}{2}\sqrt{(2x^2y^6)^2}\times\sqrt{22}$

$=\frac{1}{2}\times 2x^2y^6\times\sqrt{22}$

$=x^2y^6\sqrt{22}$

d $\frac{3}{5}\sqrt{80x^{18}y^{16}}$

$=\frac{3}{5}\sqrt{16x^{18}y^{16}\times 5}$

$=\frac{3}{5}\sqrt{(4x^9y^8)^2}\times\sqrt{5}$

$=\frac{3}{5}\times 4x^9y^8\times\sqrt{5}$

$=\frac{12x^9y^8\sqrt{5}}{5}$

e $\frac{1}{15}\sqrt{75x^4y^6z^8}$

$=\frac{1}{15}\sqrt{25x^4y^6z^8\times 3}$

$=\frac{1}{15}\sqrt{(5x^2y^3z^4)^2}\times\sqrt{3}$

$=\frac{1}{15}\times 5x^2y^3z^4\times\sqrt{3}$

$=\frac{x^2y^3z^4\sqrt{3}}{3}$

f $\frac{2}{3}\sqrt{54x^8y^2z^4}$

$=\frac{2}{3}\sqrt{9x^8y^2z^4\times 6}$

$=\frac{2}{3}\sqrt{(3x^4yz^2)^2}\times\sqrt{6}$

$=\frac{2}{3}\times 3x^4yz^2\times\sqrt{6}$

$=2x^4yz^2\sqrt{6}$

g $\frac{3}{2}\sqrt{48x^9y^3z^5}$

$=\frac{3}{2}\sqrt{16x^8y^2z^4\times 3xyz}$

$=\frac{3}{2}\sqrt{(4x^4yz^2)^2}\times\sqrt{3xyz}$

$=\frac{3}{2}\times 4x^4yz^2\times\sqrt{3xyz}$

$=6x^4yz^2\sqrt{3xyz}$

h $\frac{5}{2}\sqrt{108x^7yz^5}$

$=\frac{5}{2}\sqrt{36x^6z^4\times 3xyz}$

$=\frac{5}{2}\sqrt{(6x^3z^2)^2}\times\sqrt{3xyz}$

$=\frac{5}{2}\times 6x^3z^2\times\sqrt{3xyz}$

$=15x^3z^2\sqrt{3xyz}$

Adding and Subtracting Surds

Multiple Choice

1 **Option A**

$3\sqrt{5} + 2\sqrt{5}$

$= 5\sqrt{5}$

2 **Option C**

$4\sqrt{7} - \sqrt{7}$

$= 3\sqrt{7}$

3 **Option D**

$\sqrt{12} + \sqrt{27}$

$= \sqrt{4 \times 3} + \sqrt{9 \times 3}$

$= 2\sqrt{3} + 3\sqrt{3}$

$= 5\sqrt{3}$

4 **Option A**

$4\sqrt{50} - 5\sqrt{8}$

$= 4\sqrt{25 \times 2} - 5\sqrt{4 \times 2}$

$= 4 \times 5\sqrt{2} - 5 \times 2\sqrt{2}$

$= 20\sqrt{2} - 10\sqrt{2}$

$= 10\sqrt{2}$

5 **Option C**

$\sqrt{18x} - 2\sqrt{8x}$

$= \sqrt{9 \times 2x} - 2\sqrt{4 \times 2x}$

$= 3\sqrt{2x} - 2 \times 2\sqrt{2x}$

$= 3\sqrt{2x} - 4\sqrt{2x}$

$= -\sqrt{2x}$

Time To Practice

1 **a** $4\sqrt{7} + 3\sqrt{7}$

$= (4 + 3)\sqrt{7}$
$= 7\sqrt{7}$

b $5\sqrt{3} + 8\sqrt{3}$

$= (5 + 8)\sqrt{3}$
$= 13\sqrt{3}$

c $12\sqrt{2} + 3\sqrt{2}$

$= (12 + 3)\sqrt{2}$
$= 15\sqrt{2}$

d $6\sqrt{3} - 2\sqrt{3}$

$= (6 - 2)\sqrt{3}$
$= 4\sqrt{3}$

e $11\sqrt{10} - 6\sqrt{10}$

$= (11 - 6)\sqrt{10}$
$= 5\sqrt{10}$

f $15\sqrt{5} - \sqrt{5}$

$= (15 - 1)\sqrt{5}$
$= 14\sqrt{5}$

g $-2\sqrt{13} - \sqrt{13} + 4\sqrt{13}$

$= (-2 - 1 + 4)\sqrt{13}$
$= \sqrt{13}$

h $-\sqrt{21} + 5\sqrt{21} - 2\sqrt{21}$

$= (-1 + 5 - 2)\sqrt{21}$
$= 2\sqrt{21}$

2 **a** $8\sqrt{5} + 3\sqrt{3} + 7\sqrt{5} + 2\sqrt{3}$

$= (8 + 7)\sqrt{5} + (3 + 2)\sqrt{3}$
$= 15\sqrt{5} + 5\sqrt{3}$

b $12\sqrt{3} - 8\sqrt{7} + 5\sqrt{3} - 10\sqrt{7}$

$= (12 + 5)\sqrt{3} + (-8 - 10)\sqrt{7}$
$= 17\sqrt{3} - 18\sqrt{7}$

c $2\sqrt{2} - 4\sqrt{10} - 5\sqrt{2} + \sqrt{10}$

$= (2 - 5)\sqrt{2} + (-4 + 1)\sqrt{10}$
$= -3\sqrt{2} - 3\sqrt{10}$

d $-4\sqrt{11} - 2\sqrt{15} + 7\sqrt{11} + 4\sqrt{15}$

$= (-4 + 7)\sqrt{11} + (-2 + 4)\sqrt{15}$
$= 3\sqrt{11} + 2\sqrt{15}$

3 **a** $\sqrt{8} + 3\sqrt{2}$

$= \sqrt{4 \times 2} + 3\sqrt{2}$

$= 2\sqrt{2} + 3\sqrt{2}$

$= 5\sqrt{2}$

b $\sqrt{20} - \sqrt{5}$

$= \sqrt{4 \times 5} - \sqrt{5}$

$= 2\sqrt{5} - \sqrt{5}$

$= \sqrt{5}$

c $2\sqrt{27} + 4\sqrt{3}$

$= 2\sqrt{9 \times 3} + 4\sqrt{3}$

$= 2 \times 3\sqrt{3} + 4\sqrt{3}$

$= 6\sqrt{3} + 4\sqrt{3}$

$= 10\sqrt{3}$

d $2\sqrt{24} + 3\sqrt{6}$

$= 2\sqrt{4 \times 6} + 3\sqrt{6}$

$= 2 \times 2\sqrt{6} + 3\sqrt{6}$

$= 4\sqrt{6} + 3\sqrt{6}$

$= 7\sqrt{6}$

e $3\sqrt{72} + 2\sqrt{98}$

$= 3\sqrt{36 \times 2} + 3\sqrt{49 \times 2}$

$= 3 \times 6\sqrt{2} + 2 \times 7\sqrt{2}$

$= 18\sqrt{2} + 14\sqrt{2}$

$= 32\sqrt{2}$

f $2\sqrt{125} - 3\sqrt{45}$

$= 2\sqrt{25 \times 5} - 3\sqrt{9 \times 5}$

$= 2 \times 5\sqrt{5} - 3 \times 3\sqrt{5}$

$= 10\sqrt{5} - 9\sqrt{5}$

$= \sqrt{5}$

g $2\sqrt{80} - \sqrt{45} + 2\sqrt{63}$

$= 2\sqrt{16 \times 5} - \sqrt{9 \times 5} + 2\sqrt{9 \times 7}$

$= 2 \times 4\sqrt{5} - 3\sqrt{5} + 2 \times 3\sqrt{7}$

$= 8\sqrt{5} - 3\sqrt{5} + 6\sqrt{7}$

$= 5\sqrt{5} + 6\sqrt{7}$

h $\sqrt{36} - \sqrt{108} - 3\sqrt{3}$

$= 6 - \sqrt{36 \times 3} - 3\sqrt{3}$

$= 6 - 6\sqrt{3} - 3\sqrt{3}$

$= 6 - 9\sqrt{3}$

4 a $2\sqrt{x} + 5\sqrt{y} + 6\sqrt{x} - 2\sqrt{y}$

$= 8\sqrt{x} + 3\sqrt{y}$

b $7\sqrt{x} - \sqrt{y} + 2\sqrt{9x} - 3\sqrt{y}$

$= 7\sqrt{x} - 4\sqrt{y} + 2 \times 3\sqrt{x}$

$= 7\sqrt{x} - 4\sqrt{y} + 6\sqrt{x}$

$= 13\sqrt{x} - 4\sqrt{y}$

c $10\sqrt{y} - 2\sqrt{4y} + 2\sqrt{x} - \sqrt{x}$

$= 10\sqrt{y} - 2 \times 2\sqrt{y} + \sqrt{x}$

$= 10\sqrt{y} - 4\sqrt{y} + \sqrt{x}$

$= 6\sqrt{y} + \sqrt{x}$

d $6\sqrt{x} - 12\sqrt{9y} - 7\sqrt{x} + 2\sqrt{y}$

$= -\sqrt{x} - 12 \times 3\sqrt{y} + 2\sqrt{y}$

$= -\sqrt{x} - 36\sqrt{y} + 2\sqrt{y}$

$= -\sqrt{x} - 34\sqrt{y}$

Challenge Yourself

1 $4\sqrt{2} + \sqrt{32}$

$= 4\sqrt{2} + \sqrt{16 \times 2}$

$= 4\sqrt{2} + 4\sqrt{2}$

$= 8\sqrt{2}$

$\therefore a = 8 \; and \; b = 2$

2 $3\sqrt{125} - 2\sqrt{45}$

$= 3\sqrt{25 \times 5} - 2\sqrt{9 \times 5}$

$= 3 \times 5\sqrt{5} - 2 \times 3\sqrt{5}$

$= 15\sqrt{5} - 6\sqrt{5}$

$= 9\sqrt{5}$

$\therefore a = 9 \; and \; b = 5$

3 $4\sqrt{50x} - 3\sqrt{18x}$

$= 4\sqrt{25 \times 2x} - 3\sqrt{9 \times 2x}$

$= 4 \times 5\sqrt{2x} - 3 \times 3\sqrt{2x}$

$= 20\sqrt{2x} - 9\sqrt{2x}$

$= 11\sqrt{2x}$

$\therefore a = 11 \; and \; b = 2$

4 **a** $10\sqrt{3a} - 15\sqrt{27a} + 8\sqrt{12a}$

$= 10\sqrt{3a} - 15\sqrt{9 \times 3a} + 8\sqrt{4 \times 3a}$

$= 10\sqrt{3a} - 15 \times 3\sqrt{3a} + 8 \times 2\sqrt{3a}$

$= 10\sqrt{3a} - 45\sqrt{3a} + 16\sqrt{3a}$

$= -19\sqrt{3a}$

b $6\sqrt{2b} + \sqrt{128b} - \sqrt{144}$

$= 6\sqrt{2b} + \sqrt{64 \times 2b} - 12$

$= 6\sqrt{2b} + 8\sqrt{2b} - 12$

$= 14\sqrt{2b} - 12$

c $\frac{1}{8}\sqrt{32a} - \frac{7}{6}\sqrt{18a} - \frac{1}{3}\sqrt{72a}$

$= \frac{1}{8}\sqrt{16 \times 2a} - \frac{7}{6}\sqrt{9 \times 2a} - \frac{1}{3}\sqrt{36 \times 2a}$

$= \frac{1}{8} \times 4\sqrt{2a} - \frac{7}{6} \times 3\sqrt{2a} - \frac{1}{3} \times 6\sqrt{2a}$

$= \frac{1}{2}\sqrt{2a} - \frac{7}{2}\sqrt{2a} - 2\sqrt{2a}$

$= -5\sqrt{2a}$

d $\frac{1}{2}\sqrt{150ab} + \frac{1}{4}\sqrt{96ab} - \frac{1}{6}\sqrt{6ab}$

$= \frac{1}{2}\sqrt{25 \times 6ab} + \frac{1}{4}\sqrt{16 \times 6ab} - \frac{1}{6}\sqrt{6ab}$

$= \frac{1}{2} \times 5\sqrt{6ab} + \frac{1}{4} \times 4\sqrt{6ab} - \frac{1}{6}\sqrt{6ab}$

$= \frac{5}{2}\sqrt{6ab} + \sqrt{6ab} - \frac{1}{6}\sqrt{6ab}$

$= \frac{10\sqrt{6ab}}{3}$

Multiplying Surds

Multiple Choice

1 **Option B**

$\sqrt{5} \times \sqrt{2}$

$= \sqrt{5 \times 2}$

$= \sqrt{10}$

2 **Option A**

$2\sqrt{18} \times \sqrt{3}$

$= 2\sqrt{54}$

$= 2\sqrt{9 \times 6}$

$= 2 \times 3 \times \sqrt{6}$

$= 6\sqrt{6}$

3 **Option A**

$3\sqrt{5} \times 2\sqrt{10}$

$= 6\sqrt{5 \times 10}$

$= 6\sqrt{50}$

$= 6\sqrt{25 \times 2}$

$= 6 \times 5\sqrt{2}$

$= 30\sqrt{2}$

4 **Option C**

$2\sqrt{45} \times \sqrt{63}$

$= 2\sqrt{9 \times 5} \times \sqrt{9 \times 7}$

$= 2 \times 3\sqrt{5} \times 3\sqrt{7}$

$= 18\sqrt{5 \times 7}$

$= 18\sqrt{35}$

5 **Option C**

$\sqrt{12x^2} \times \sqrt{3}$

$= \sqrt{12x^2 \times 3}$

$= \sqrt{36x^2}$

$= 6x$

Time To Practice

1 a $\sqrt{2} \times \sqrt{7}$

$= \sqrt{2 \times 7}$

$= \sqrt{14}$

b $\sqrt{6} \times \sqrt{7}$

$= \sqrt{6 \times 7}$

$= \sqrt{42}$

c $\sqrt{5} \times \sqrt{6}$

$= \sqrt{5 \times 6}$

$= \sqrt{30}$

d $\sqrt{10} \times \sqrt{11}$

$= \sqrt{10 \times 11}$

$= \sqrt{110}$

e $\sqrt{2} \times \sqrt{17}$

$= \sqrt{2 \times 17}$

$= \sqrt{34}$

f $3\sqrt{2} \times \sqrt{21}$

$= 3\sqrt{2 \times 21}$

$= 3\sqrt{42}$

g $2\sqrt{11} \times 5\sqrt{3}$

$= 10\sqrt{11 \times 3}$

$= 10\sqrt{33}$

h $6\sqrt{3} \times 10\sqrt{7}$

$= 60\sqrt{3 \times 7}$

$= 60\sqrt{21}$

i $3\sqrt{5} \times 4\sqrt{7}$

$= 12\sqrt{5 \times 7}$

$= 12\sqrt{35}$

2 a $2\sqrt{8} \times \sqrt{2}$

$= 2\sqrt{8 \times 2}$

$= 2\sqrt{16}$

$= 2 \times 4$

$= 8$

b $3\sqrt{18} \times \sqrt{12}$

$= 3\sqrt{9 \times 2} \times \sqrt{4 \times 3}$

$= 3 \times 3\sqrt{2} \times 2\sqrt{3}$

$= 18\sqrt{2 \times 3}$

$= 18\sqrt{6}$

2 **c** $\sqrt{12} \times 4\sqrt{48}$

$= \sqrt{4 \times 3} \times 4\sqrt{16 \times 3}$

$= 2\sqrt{3} \times 4 \times 4\sqrt{3}$

$= 2\sqrt{3} \times 16\sqrt{3}$

$= 32 \times 3$

$= 96$

d $\sqrt{32} \times \sqrt{72}$

$= \sqrt{16 \times 2} \times \sqrt{36 \times 2}$

$= 4\sqrt{2} \times 6\sqrt{2}$

$= 24 \times 2$

$= 48$

e $2\sqrt{75} \times 3\sqrt{45}$

$= 2\sqrt{25 \times 3} \times 3\sqrt{9 \times 5}$

$= 2 \times 5\sqrt{3} \times 3 \times 3\sqrt{5}$

$= 10\sqrt{3} \times 9\sqrt{5}$

$= 90\sqrt{15}$

f $\sqrt{20} \times \sqrt{125}$

$= \sqrt{4 \times 5} \times \sqrt{25 \times 5}$

$= 2\sqrt{5} \times 5\sqrt{5}$

$= 10 \times 5$

$= 50$

g $4\sqrt{63} \times 2\sqrt{6}$

$= 4\sqrt{9 \times 7} \times 2\sqrt{6}$

$= 4 \times 3\sqrt{7} \times 2\sqrt{6}$

$= 24\sqrt{7 \times 6}$

$= 24\sqrt{42}$

h $3\sqrt{50} \times 2\sqrt{20}$

$= 3\sqrt{25 \times 2} \times 2\sqrt{4 \times 5}$

$= 3 \times 5\sqrt{2} \times 2 \times 2\sqrt{5}$

$= 15\sqrt{2} \times 4\sqrt{5}$

$= 60\sqrt{2 \times 5}$

$= 60\sqrt{10}$

i $4\sqrt{15} \times 2\sqrt{18}$

$= 4\sqrt{15} \times 2 \times \sqrt{9 \times 2}$

$= 4\sqrt{15} \times 2 \times 3\sqrt{2}$

$= 4\sqrt{15} \times 6\sqrt{2}$

$= 24\sqrt{15 \times 2}$

$= 24\sqrt{30}$

3 **a** $\dfrac{\sqrt{5}}{3}+\dfrac{\sqrt{5}}{4}$

$=\dfrac{4\sqrt{5}+3\sqrt{5}}{12}$

$=\dfrac{7\sqrt{5}}{12}$

b $\dfrac{\sqrt{2}}{5}-\dfrac{\sqrt{2}}{6}$

$=\dfrac{6\sqrt{2}-5\sqrt{2}}{30}$

$=\dfrac{\sqrt{2}}{30}$

c $\dfrac{\sqrt{7}}{4}-\dfrac{\sqrt{7}}{12}$

$=\dfrac{3\sqrt{7}}{12}-\dfrac{\sqrt{7}}{12}$

$=\dfrac{2\sqrt{7}}{12}$

$=\dfrac{\sqrt{7}}{6}$

d $\dfrac{2\sqrt{2}}{5}-\dfrac{\sqrt{2}}{2}$

$=\dfrac{4\sqrt{2}-5\sqrt{2}}{10}$

$=-\dfrac{\sqrt{2}}{10}$

e $\dfrac{3\sqrt{3}}{7}-\dfrac{\sqrt{3}}{2}$

$=\dfrac{6\sqrt{3}-7\sqrt{3}}{14}$

$=-\dfrac{\sqrt{3}}{14}$

f $\dfrac{7\sqrt{7}}{4}+\dfrac{\sqrt{7}}{12}$

$=\dfrac{21\sqrt{7}}{12}+\dfrac{\sqrt{7}}{12}$

$=\dfrac{22\sqrt{7}}{12}$

$=\dfrac{11\sqrt{7}}{6}$

3 **g** $\frac{\sqrt{8}}{3} - \frac{\sqrt{2}}{5}$

$= \frac{2\sqrt{2}}{3} - \frac{\sqrt{2}}{5}$

$= \frac{10\sqrt{2} - 3\sqrt{2}}{15}$

$= \frac{7\sqrt{2}}{15}$

h $\frac{\sqrt{12}}{4} + \frac{\sqrt{3}}{6}$

$= \frac{2\sqrt{3}}{4} + \frac{\sqrt{3}}{6}$

$= \frac{12\sqrt{3} + 4\sqrt{3}}{24}$

$= \frac{16\sqrt{3}}{24}$

$= \frac{2\sqrt{3}}{3}$

i $\frac{2\sqrt{18}}{3} - \frac{\sqrt{72}}{2}$

$= \frac{2 \times 3\sqrt{2}}{3} - \frac{6\sqrt{2}}{2}$

$= \frac{6\sqrt{2}}{3} - \frac{6\sqrt{2}}{2}$

$= \frac{12\sqrt{2} - 18\sqrt{2}}{6}$

$= \frac{-6\sqrt{2}}{6}$

$= -\sqrt{2}$

4 **a** $\sqrt{2x^2} \times \sqrt{3}$

$= \sqrt{2x^2 \times 3}$

$= \sqrt{6x^2}$

$= x\sqrt{6}$

b $\sqrt{8x^2y} \times \sqrt{y}$

$= \sqrt{8x^2y \times y} = \sqrt{8x^2y^2}$

$= \sqrt{4x^2y^2 \times 2}$

$= \sqrt{(2xy)^2 \times 2}$

$= 2xy\sqrt{2}$

4 c $\sqrt{2x^2y^2} \times 3\sqrt{27}$

$= xy\sqrt{2} \times 3\sqrt{9 \times 3}$

$= xy\sqrt{2} \times 3 \times 3\sqrt{3}$

$= xy\sqrt{2} \times 9\sqrt{3}$

$= 9xy\sqrt{6}$

d $\sqrt{12x^8y} \times 6\sqrt{2y}$

$= 6\sqrt{12x^8y \times 2y}$

$= 6\sqrt{24x^8y^2}$

$= 6\sqrt{4x^8y^2 \times 6}$

$= 6\sqrt{(2x^4y)^2 \times 6}$

$= 6 \times 2x^4y\sqrt{6}$

$= 12x^4y\sqrt{6}$

e $\sqrt{2x^6y^2} \times \sqrt{24x^4y^8}$

$= \sqrt{2x^6y^2 \times 24x^4y^8}$

$= \sqrt{48x^{10}y^{10}}$

$= \sqrt{16x^{10}y^{10} \times 3}$

$= \sqrt{(4x^5y^5)^2 \times 3}$

$= 4x^5y^5\sqrt{3}$

f $3\sqrt{7x^2y^3} \times 2\sqrt{x^2y^5}$

$= 6\sqrt{7x^2y^3 \times x^2y^5}$

$= 6\sqrt{7x^4y^8}$

$= 6\sqrt{7 \times x^4y^8}$

$= 6\sqrt{7 \times (x^2y^4)^2}$

$= 6x^2y^4\sqrt{7}$

Challenge Yourself

1 a $\frac{1}{2}\sqrt{60x^4} \times \sqrt{18x^2}$

$= \frac{1}{2}\sqrt{4x^4 \times 15} \times \sqrt{9x^2 \times 2}$

$= \frac{1}{2}\sqrt{(2x^2)^2 \times 15} \times \sqrt{(3x)^2 \times 2}$

$= \frac{1}{2} \times 2x^2\sqrt{15} \times 3x\sqrt{2}$

$= 3x^3\sqrt{15 \times 2}$

$= 3x^3\sqrt{30}$

b $\frac{1}{4}\sqrt{128x^8} \times 2\sqrt{75x^2}$

$= \frac{1}{4}\sqrt{64x^8 \times 2} \times 2\sqrt{25x^2 \times 3}$

$= \frac{1}{4}\sqrt{(8x^4)^2 \times 2} \times 2\sqrt{(5x)^2 \times 3}$

$= \frac{1}{4} \times 8x^4\sqrt{2} \times 2 \times 5x\sqrt{3}$

$= 2x^4\sqrt{2} \times 10x\sqrt{3}$

$= 20x^5\sqrt{6}$

1 c $\frac{2}{3}\sqrt{12x^8y^2} \times 6\sqrt{32y^6}$

$= \frac{2}{3}\sqrt{4x^8y^2 \times 3} \times 6\sqrt{16y^6 \times 2}$

$= \frac{2}{3}\sqrt{(2x^4y)^2 \times 3} \times 6\sqrt{(4y^3)^2 \times 2}$

$= \frac{2}{3} \times 2x^4y\sqrt{3} \times 6 \times 4y^3\sqrt{2}$

$= \frac{4x^4y\sqrt{3}}{3} \times 24y^3\sqrt{2}$

$= 4x^4y\sqrt{3} \times 8y^3\sqrt{2}$

$= 32x^4y^4\sqrt{3 \times 2}$

$= 32x^4y^4\sqrt{6}$

d $\frac{3}{8}\sqrt{63x^2y^2} \times \frac{6}{7}\sqrt{96y^{12}}$

$= \frac{3}{8}\sqrt{9x^2y^2 \times 7} \times \frac{6}{7}\sqrt{16y^{12} \times 6}$

$= \frac{3}{8}\sqrt{(3xy)^2 \times 7} \times \frac{6}{7}\sqrt{(4y^6)^2 \times 6}$

$= \frac{3}{8} \times 3xy\sqrt{7} \times \frac{6}{7} \times 4y^6\sqrt{6}$

$= \frac{9xy\sqrt{7}}{8} \times \frac{24y^6\sqrt{6}}{7}$

$= \frac{9xy\sqrt{7}}{1} \times \frac{3y^6\sqrt{6}}{7}$

$= \frac{27xy^7\sqrt{7 \times 6}}{7}$

$= \frac{27xy^7\sqrt{42}}{7}$

1 e $\frac{2}{3}\sqrt{88x^4y^2} \times \frac{4}{5}\sqrt{48y^6}$

$= \frac{2}{3}\sqrt{4x^4y^2 \times 22} \times \frac{4}{5}\sqrt{16y^6 \times 3}$

$= \frac{2}{3}\sqrt{(2x^2y)^2 \times 22} \times \frac{4}{5}\sqrt{(4y^3)^2 \times 3}$

$= \frac{2}{3} \times 2x^2y\sqrt{22} \times \frac{4}{5} \times 4y^3\sqrt{3}$

$= \frac{4x^2y\sqrt{22}}{3} \times \frac{16y^3\sqrt{3}}{5}$

$= \frac{64x^2y^4\sqrt{22 \times 3}}{15}$

$= \frac{64x^2y^4\sqrt{66}}{15}$

f $\frac{1}{6}\sqrt{200x^6y^4} \times \frac{3}{7}\sqrt{60x^8y^2}$

$= \frac{1}{6}\sqrt{100x^6y^4 \times 2} \times \frac{3}{7}\sqrt{4x^8y^2 \times 15}$

$= \frac{1}{6}\sqrt{(10x^3y^2)^2 \times 2} \times \frac{3}{7}\sqrt{(2x^4y)^2 \times 15}$

$= \frac{1}{6} \times 10x^3y^2\sqrt{2} \times \frac{3}{7} \times 2x^4y\sqrt{15}$

$= \frac{10x^3y^2\sqrt{2}}{6} \times \frac{6x^4y\sqrt{15}}{7}$

$= \frac{10x^7y^3\sqrt{2 \times 15}}{7}$

$= \frac{60x^7y^3\sqrt{30}}{7}$

Dividing Surds

Multiple Choice

1 **Option C**

$\sqrt{48} \div \sqrt{12}$

$= \sqrt{\frac{48}{12}}$

$= \sqrt{4}$

$= 2$

2 **Option A**

$\sqrt{18} \div \sqrt{2}$

$= \sqrt{\frac{18}{2}}$

$= \sqrt{9}$

$= 3$

3 **Option B**

$\sqrt{75} \div \sqrt{5}$

$= \sqrt{\frac{75}{5}}$

$= \sqrt{15}$

4 **Option C**

$3\sqrt{27} \div \sqrt{3}$

$= 3\sqrt{\frac{27}{9}}$

$= 3\sqrt{3}$

5 **Option D**

$\sqrt{72x^2} \div \sqrt{8}$

$= \sqrt{\frac{72x^2}{8}}$

$= \sqrt{9x^2}$

$= \sqrt{(3x)^2}$

$= 3x$

Time To Practice

1 **a** $\sqrt{16} \div \sqrt{4}$

$= \sqrt{\frac{16}{4}}$

$= \sqrt{4} = 2$

b $\sqrt{25} \div \sqrt{5}$

$= \sqrt{\frac{25}{5}}$

$= \sqrt{5}$

c $\sqrt{81} \div \sqrt{9}$

$= \sqrt{\frac{81}{9}}$

$= \sqrt{9}$

$= 3$

d $\sqrt{64} \div \sqrt{2}$

$= \sqrt{\frac{64}{2}} = \sqrt{32}$

$= \sqrt{16 \times 2}$

$= 4\sqrt{2}$

e $3\sqrt{36} \div 2\sqrt{4}$

$= \frac{3}{2}\sqrt{\frac{36}{4}}$

$= \frac{3}{2}\sqrt{9}$

$= \frac{3}{2} \times 3$

$= \frac{9}{2}$

f $2\sqrt{45} \div \sqrt{5}$

$= 2\sqrt{\frac{45}{5}}$

$= 2\sqrt{9}$

$= 2 \times 3$

$= 6$

g $4\sqrt{72} \div 2\sqrt{18}$

$= \frac{4}{2}\sqrt{\frac{72}{18}}$

$= 2\sqrt{4}$

$= 2 \times 2$

$= 4$

h $\sqrt{200} \div 2\sqrt{5}$

$= \frac{1}{2}\sqrt{\frac{200}{5}}$

$= \frac{1}{2}\sqrt{40} = \frac{1}{2}\sqrt{4 \times 10}$

$= \frac{1}{2} \times 2\sqrt{10}$

$= \sqrt{10}$

1 i $8\sqrt{18} \div 4\sqrt{2}$

$= \frac{8}{4}\sqrt{\frac{18}{2}}$

$= 2\sqrt{9}$

$= 2 \times 3$

$= 6$

j $\sqrt{\frac{9}{64}}$

$= \frac{\sqrt{9}}{\sqrt{64}}$

$= \frac{3}{8}$

k $\sqrt{\frac{12}{25}}$

$= \frac{\sqrt{12}}{\sqrt{25}}$

$= \frac{\sqrt{4 \times 3}}{5}$

$= \frac{2\sqrt{3}}{5}$

l $3\sqrt{\frac{27}{49}}$

$= \frac{3\sqrt{27}}{\sqrt{49}}$

$= \frac{3\sqrt{9 \times 3}}{7}$

$= \frac{3 \times 3\sqrt{3}}{7}$

$= \frac{9\sqrt{3}}{7}$

2 a $\frac{3\sqrt{64}}{5\sqrt{8}}$

$= \frac{3}{5}\sqrt{\frac{64}{8}} = \frac{3}{5}\sqrt{8}$

$= \frac{3}{5}\sqrt{4 \times 3}$

$= \frac{3}{5} \times 2\sqrt{2}$

$= \frac{6\sqrt{2}}{5}$

b $\frac{\sqrt{60}}{5\sqrt{5}}$

$= \frac{1}{5}\sqrt{\frac{60}{5}} = \frac{1}{5}\sqrt{12}$

$= \frac{1}{5}\sqrt{4 \times 3}$

$= \frac{1}{5} \times 2\sqrt{3}$

$= \frac{2\sqrt{3}}{5}$

2 **c** $\dfrac{9\sqrt{80}}{12\sqrt{4}}$

$= \dfrac{3}{4}\sqrt{\dfrac{80}{4}}$

$= \dfrac{3}{4}\sqrt{20}$

$= \dfrac{3}{4}\sqrt{4 \times 5}$

$= \dfrac{3}{4} \times 2\sqrt{5}$

$= \dfrac{3\sqrt{5}}{2}$

d $\dfrac{3\sqrt{88}}{14\sqrt{11}}$

$= \dfrac{3}{14}\sqrt{\dfrac{88}{11}}$

$= \dfrac{3}{14}\sqrt{8}$

$= \dfrac{3}{14}\sqrt{4 \times 2}$

$= \dfrac{3}{14} \times 2\sqrt{2}$

$= \dfrac{3\sqrt{2}}{7}$

e $\dfrac{3\sqrt{108}}{11\sqrt{6}}$

$= \dfrac{3}{11}\sqrt{\dfrac{108}{6}}$

$= \dfrac{3}{11}\sqrt{18}$

$= \dfrac{3}{11}\sqrt{9 \times 2}$

$= \dfrac{3}{11} \times 3\sqrt{2}$

$= \dfrac{9\sqrt{2}}{11}$

f $\dfrac{12\sqrt{56}}{18\sqrt{2}}$

$= \dfrac{2}{3}\sqrt{\dfrac{56}{2}}$

$= \dfrac{2}{3}\sqrt{28}$

$= \dfrac{2}{3}\sqrt{4 \times 7}$

$= \dfrac{2}{3} \times 2\sqrt{7}$

$= \dfrac{4\sqrt{7}}{3}$

3 a $\frac{1}{\sqrt{2}} \times \frac{\sqrt{6}}{3}$

$= \frac{\sqrt{6}}{3\sqrt{2}}$

$= \frac{1}{3}\sqrt{\frac{6}{2}}$

$= \frac{1}{3}\sqrt{3}$

$= \frac{\sqrt{3}}{3}$

b $\frac{\sqrt{10}}{\sqrt{3}} \times \frac{\sqrt{6}}{2}$

$= \frac{\sqrt{60}}{2\sqrt{3}}$

$= \frac{1}{2}\sqrt{\frac{60}{3}}$

$= \frac{1}{2}\sqrt{20}$

$= \frac{1}{2}\sqrt{4 \times 5}$

$= \frac{1}{2} \times 2\sqrt{5}$

$= \sqrt{5}$

c $\frac{\sqrt{12}}{3} \times \frac{\sqrt{9}}{2\sqrt{6}}$

$= \frac{\sqrt{108}}{6\sqrt{6}}$

$= \frac{1}{6}\sqrt{\frac{108}{6}} = \frac{1}{6}\sqrt{18}$

$= \frac{1}{6}\sqrt{9 \times 2}$

$= \frac{1}{6} \times 3\sqrt{2}$

$= \frac{\sqrt{2}}{2}$

d $\frac{\sqrt{5}}{2} \div \frac{4}{\sqrt{10}}$

$= \frac{\sqrt{5}}{2} \times \frac{\sqrt{10}}{4}$

$= \frac{\sqrt{50}}{8}$

$= \frac{\sqrt{25 \times 2}}{8}$

$= \frac{5\sqrt{2}}{8}$

3 **e** $\dfrac{\sqrt{18}}{\sqrt{2}} \div \dfrac{3}{\sqrt{6}}$

$= \sqrt{\dfrac{18}{2}} \times \dfrac{\sqrt{6}}{3}$

$= \sqrt{9} \times \dfrac{\sqrt{6}}{3}$

$= 3 \times \dfrac{\sqrt{6}}{3}$

$= \sqrt{6}$

f $\dfrac{5}{\sqrt{6}} \div \dfrac{2}{9\sqrt{48}}$

$= \dfrac{5}{\sqrt{6}} \times \dfrac{9\sqrt{48}}{2}$

$= \dfrac{45\sqrt{48}}{2\sqrt{6}}$

$= \dfrac{45}{2}\sqrt{\dfrac{48}{6}}$

$= \dfrac{45}{2}\sqrt{8}$

$= \dfrac{45}{2}\sqrt{4 \times 2}$

$= \dfrac{45}{2} \times 2\sqrt{2}$

$= 45\sqrt{2}$

4 **a** $\dfrac{\sqrt{x^8y^4}}{\sqrt{x^2y^2}}$

$= \dfrac{\sqrt{(x^4y^2)^2}}{\sqrt{(xy)^2}}$

$= \dfrac{x^4y^2}{xy}$

$= x^3y$

b $\dfrac{\sqrt{8x^8y^4}}{\sqrt{2x^4y^6}}$

$= \sqrt{\dfrac{8x^8y^4}{2x^4y^6}}$

$= \sqrt{\dfrac{4x^4}{y^2}} = \sqrt{\left(\dfrac{2x^2}{y}\right)^2}$

$= \dfrac{2x^2}{y}$

4 **c** $\dfrac{\sqrt{9x^2}}{\sqrt{x^4y^8}}$

$= \dfrac{\sqrt{(3x)^2}}{\sqrt{(x^2y^4)^2}}$

$= \dfrac{3x}{x^2y^4}$

$= \dfrac{3}{xy^4}$

d $\dfrac{\sqrt{x^4y^2}}{\sqrt{x^6y^4}} \div \dfrac{\sqrt{x^8}}{\sqrt{y^6}}$

$= \dfrac{\sqrt{(x^2y)^2}}{\sqrt{(x^3y^2)^2}} \div \dfrac{\sqrt{(x^4)^2}}{\sqrt{(y^3)^2}}$

$= \dfrac{x^2y}{x^3y^2} \div \dfrac{x^4}{y^3}$

$= \dfrac{1}{xy} \times \dfrac{y^3}{x^4}$

$= \dfrac{1}{x} \times \dfrac{y^2}{x^4} = \dfrac{y^2}{x^5}$

e $\dfrac{\sqrt{xy}}{\sqrt{x^5y^7}} \div \dfrac{\sqrt{x^2y^4}}{\sqrt{12x^8y^{12}}}$

$= \sqrt{\dfrac{xy}{x^5y^7}} \div \dfrac{\sqrt{(xy^2)^2}}{\sqrt{4x^8y^{12} \times 3}}$

$= \sqrt{\dfrac{1}{x^4y^6}} \div \dfrac{xy^2}{\sqrt{(2x^4y^6)^2 \times 3}}$

$= \dfrac{1}{\sqrt{(x^2y^3)^2}} \div \dfrac{xy^2}{2x^4y^6\sqrt{3}}$

$= \dfrac{1}{x^2y^3} \times \dfrac{2x^4y^6\sqrt{3}}{xy^2}$

$= 2xy\sqrt{3}$

f $\dfrac{2\sqrt{6x^5y^3}}{\sqrt{x^3y}} \div \dfrac{3\sqrt{y}}{\sqrt{y^7}}$

$= 2\sqrt{\dfrac{6x^5y^3}{x^3y}} \div 3\sqrt{\dfrac{y}{y^7}}$

$= 2\sqrt{6x^2y^2} \div 3\sqrt{\dfrac{1}{y^6}}$

$= 2\sqrt{6(xy)^2} \div 3 \times \dfrac{1}{\sqrt{(y^3)^2}}$

$= 2xy\sqrt{6} \div \dfrac{3}{y^3}$

$= 2xy\sqrt{6} \times \dfrac{y^3}{3}$

$= \dfrac{2xy^4\sqrt{6}}{3}$

Challenge Yourself

1 a

$$\frac{9\sqrt{12} \times 2\sqrt{18}}{3\sqrt{8}}$$

$$= \frac{9\sqrt{4 \times 3} \times 2\sqrt{9 \times 2}}{3\sqrt{4 \times 2}}$$

$$= \frac{18\sqrt{3} \times 6\sqrt{2}}{6\sqrt{2}}$$

$$= 18\sqrt{3}$$

b

$$\left(\frac{\left(5\sqrt{3}\right)^2 \times 2\sqrt{5}}{3}\right) \times \frac{\sqrt{3}}{6\sqrt{5}}$$

$$= \frac{25 \times 3 \times 2\sqrt{5}}{3} \times \frac{\sqrt{3}}{6\sqrt{5}}$$

$$= \frac{50\sqrt{5}}{1} \times \frac{\sqrt{3}}{6\sqrt{5}}$$

$$= \frac{25}{1} \times \frac{\sqrt{3}}{3}$$

$$= \frac{25\sqrt{3}}{3}$$

c

$$\frac{16}{\sqrt{3}} \div \left(\frac{\left(2\sqrt{2}\right)^2 \times \left(2\sqrt{5}\right)^2}{5\sqrt{60}}\right)$$

$$= \frac{16}{\sqrt{3}} \div \left(\frac{4 \times 2 \times 4 \times 5}{5\sqrt{4 \times 15}}\right)$$

$$= \frac{16}{\sqrt{3}} \div \frac{160}{10\sqrt{15}}$$

$$= \frac{16}{\sqrt{3}} \times \frac{10\sqrt{15}}{160}$$

$$= \frac{1}{\sqrt{3}} \times \frac{\sqrt{15}}{1}$$

$$= \frac{1}{1} \times \frac{\sqrt{15}}{\sqrt{3}}$$

$$= \sqrt{\frac{15}{3}}$$

$$= \sqrt{5}$$

d

$$\left(\frac{\sqrt{54} \times 3\sqrt{22}}{3\sqrt{8}}\right) \div \frac{16\sqrt{33}}{\sqrt{3}}$$

$$= \left(\frac{\sqrt{9 \times 6} \times 3\sqrt{22}}{3\sqrt{4 \times 2}}\right) \times \frac{\sqrt{3}}{16\sqrt{33}}$$

$$= \left(\frac{3\sqrt{6} \times 3\sqrt{22}}{6\sqrt{2}}\right) \times \frac{1}{16}\sqrt{\frac{3}{33}}$$

$$= \left(\frac{9\sqrt{6 \times 22}}{6\sqrt{2}}\right) \times \frac{1}{16}\sqrt{\frac{1}{11}}$$

$$= \left(\frac{3}{2}\sqrt{\frac{132}{2}}\right) \times \frac{1}{16\sqrt{11}}$$

$$= \frac{3\sqrt{66}}{2} \times \frac{1}{16\sqrt{11}}$$

$$= \frac{3\sqrt{66}}{32\sqrt{11}} = \frac{3}{32}\sqrt{\frac{66}{11}}$$

$$= \frac{3\sqrt{6}}{32}$$

2 **a** $\dfrac{\sqrt{x^8y^{12}}}{\sqrt{3x^2y^3}} \div \dfrac{4\sqrt{x^5y^7}}{\sqrt{21x}}$

$= \dfrac{\sqrt{x^8y^{12}}}{\sqrt{3x^2y^3}} \times \dfrac{\sqrt{21x}}{4\sqrt{x^5y^7}}$

$= \dfrac{\sqrt{x^8y^{12} \times 21x}}{4\sqrt{3x^2y^3 \times x^5y^7}}$

$= \dfrac{\sqrt{21x^9y^{12}}}{4\sqrt{3x^7y^{10}}}$

$= \dfrac{1}{4}\sqrt{\dfrac{21x^9y^{12}}{3x^7y^{10}}}$

$= \dfrac{1}{4}\sqrt{7x^2y^2}$

$= \dfrac{1}{4}\sqrt{7(xy)^2}$

$= \dfrac{xy\sqrt{7}}{4}$

b $\dfrac{2\sqrt{5x^4y^2}}{\sqrt{4x^3y^5}} \div \dfrac{3\sqrt{10x^7y}}{\sqrt{6y^6}}$

$= \dfrac{2\sqrt{5x^4y^2}}{\sqrt{4x^3y^5}} \times \dfrac{\sqrt{6y^6}}{3\sqrt{10x^7y}}$

$= \dfrac{2\sqrt{5x^4y^2 \times 6y^6}}{3\sqrt{4x^3y^5 \times 10x^7y}}$

$= \dfrac{2}{3}\sqrt{\dfrac{30x^4y^8}{40x^{10}y^6}}$

$= \dfrac{2}{3}\sqrt{\dfrac{3y^2}{4x^6}}$

$= \dfrac{2}{3} \times \dfrac{\sqrt{3y^2}}{\sqrt{(2x^3)^2}}$

$= \dfrac{2}{3} \times \dfrac{y\sqrt{3}}{2x^3}$

$= \dfrac{y\sqrt{3}}{3x^3}$

Rationalizing The Denominator

Multiple Choice

1 **Option D**

$$\frac{1}{\sqrt{2}} \times \frac{\sqrt{2}}{\sqrt{2}}$$

$$= \frac{\sqrt{2}}{2}$$

2 **Option B**

$$\frac{4}{\sqrt{3}} \times \frac{\sqrt{3}}{\sqrt{3}}$$

$$= \frac{4\sqrt{3}}{3}$$

3 **Option B**

$$\frac{3}{\sqrt{12}}$$

$$= \frac{3}{\sqrt{4 \times 3}}$$

$$= \frac{3}{2\sqrt{3}}$$

$$\therefore \sqrt{3}$$

4 **Option A**

$$\frac{10}{\sqrt{18}}$$

$$= \frac{10}{\sqrt{9 \times 2}} = \frac{10}{3\sqrt{2}}$$

$$= \frac{10}{3\sqrt{2}} \times \frac{\sqrt{2}}{\sqrt{2}}$$

$$= \frac{10\sqrt{2}}{3 \times 2} = \frac{10\sqrt{2}}{6}$$

$$= \frac{5\sqrt{2}}{3}$$

5 **Option C**

$$\frac{\sqrt{20}}{\sqrt{50}}$$

$$= \frac{\sqrt{4 \times 5}}{\sqrt{25 \times 2}} = \frac{2\sqrt{5}}{5\sqrt{2}}$$

$$= \frac{2\sqrt{5}}{5\sqrt{2}} \times \frac{\sqrt{2}}{\sqrt{2}}$$

$$= \frac{2\sqrt{10}}{5 \times 2}$$

$$= \frac{2\sqrt{10}}{10}$$

$$= \frac{\sqrt{10}}{5}$$

Time To Practice

1 **a** $\frac{1}{\sqrt{7}}$

$= \frac{1}{\sqrt{7}} \times \frac{\sqrt{7}}{\sqrt{7}}$

$= \frac{\sqrt{7}}{\left(\sqrt{7}\right)^2}$

$= \frac{\sqrt{7}}{7}$

b $\frac{4}{\sqrt{5}}$

$= \frac{4}{\sqrt{5}} \times \frac{\sqrt{5}}{\sqrt{5}}$

$= \frac{4\sqrt{7}}{\left(\sqrt{5}\right)^2}$

$= \frac{4\sqrt{5}}{5}$

c $\frac{8}{\sqrt{2}}$

$= \frac{8}{\sqrt{2}} \times \frac{\sqrt{2}}{\sqrt{2}}$

$= \frac{8\sqrt{2}}{\left(\sqrt{2}\right)^2}$

$= \frac{8\sqrt{2}}{2}$

$= 4\sqrt{2}$

d $\frac{\sqrt{5}}{\sqrt{3}}$

$= \frac{\sqrt{5}}{\sqrt{3}} \times \frac{\sqrt{3}}{\sqrt{3}}$

$= \frac{\sqrt{5 \times 3}}{\left(\sqrt{3}\right)^2}$

$= \frac{\sqrt{15}}{3}$

e $\frac{\sqrt{7}}{\sqrt{2}}$

$= \frac{\sqrt{7}}{\sqrt{2}} \times \frac{\sqrt{2}}{\sqrt{2}}$

$= \frac{\sqrt{7 \times 2}}{\left(\sqrt{2}\right)^2}$

$= \frac{\sqrt{14}}{2}$

f $\frac{\sqrt{11}}{\sqrt{2}}$

$= \frac{\sqrt{11}}{\sqrt{2}} \times \frac{\sqrt{2}}{\sqrt{2}}$

$= \frac{\sqrt{11 \times 2}}{\left(\sqrt{2}\right)^2}$

$= \frac{\sqrt{22}}{2}$

2 a $\frac{4\sqrt{2}}{\sqrt{7}}$

$= \frac{4\sqrt{2}}{\sqrt{7}} \times \frac{\sqrt{7}}{\sqrt{7}}$

$= \frac{4\sqrt{14}}{(\sqrt{7})^2}$

$= \frac{4\sqrt{14}}{7}$

b $\frac{5\sqrt{2}}{\sqrt{3}}$

$= \frac{5\sqrt{2}}{\sqrt{3}} \times \frac{\sqrt{3}}{\sqrt{3}}$

$= \frac{5\sqrt{6}}{(\sqrt{3})^2}$

$= \frac{5\sqrt{6}}{3}$

c $\frac{3\sqrt{6}}{\sqrt{7}}$

$= \frac{3\sqrt{6}}{\sqrt{7}} \times \frac{\sqrt{7}}{\sqrt{7}}$

$= \frac{3\sqrt{42}}{(\sqrt{7})^2}$

$= \frac{3\sqrt{42}}{7}$

d $\frac{2\sqrt{7}}{5\sqrt{10}}$

$= \frac{2\sqrt{7}}{5\sqrt{10}} \times \frac{\sqrt{10}}{\sqrt{10}}$

$= \frac{2\sqrt{70}}{5(\sqrt{10})^2}$

$= \frac{2\sqrt{70}}{5 \times 10}$

$= \frac{\sqrt{70}}{25}$

e $\frac{4\sqrt{7}}{5\sqrt{3}}$

$= \frac{4\sqrt{7}}{5\sqrt{3}} \times \frac{\sqrt{3}}{\sqrt{3}}$

$= \frac{4\sqrt{21}}{5(\sqrt{3})^2}$

$= \frac{4\sqrt{21}}{5 \times 3}$

$= \frac{4\sqrt{21}}{15}$

f $\frac{4\sqrt{3}}{3\sqrt{5}}$

$= \frac{4\sqrt{3}}{3\sqrt{5}} \times \frac{\sqrt{5}}{\sqrt{5}}$

$= \frac{4\sqrt{15}}{3(\sqrt{5})^2}$

$= \frac{4\sqrt{15}}{3 \times 5}$

$= \frac{4\sqrt{15}}{15}$

2 g $\dfrac{8\sqrt{3}}{7\sqrt{7}}$

$= \dfrac{8\sqrt{3}}{7\sqrt{7}} \times \dfrac{\sqrt{7}}{\sqrt{7}}$

$= \dfrac{8\sqrt{21}}{7\left(\sqrt{7}\right)^2}$

$= \dfrac{8\sqrt{21}}{7 \times 7}$

$= \dfrac{8\sqrt{21}}{49}$

h $\dfrac{9\sqrt{6}}{2\sqrt{3}}$

$= \dfrac{9\sqrt{6}}{2\sqrt{3}} \times \dfrac{\sqrt{3}}{\sqrt{3}}$

$= \dfrac{9\sqrt{18}}{2 \times 3}$

$= \dfrac{9\sqrt{9 \times 2}}{6}$

$= \dfrac{9 \times 3\sqrt{2}}{6}$

$= \dfrac{9\sqrt{2}}{2}$

i $\dfrac{5\sqrt{12}}{9\sqrt{6}}$

$= \dfrac{5\sqrt{12}}{9\sqrt{6}} \times \dfrac{\sqrt{6}}{\sqrt{6}}$

$= \dfrac{5\sqrt{72}}{9 \times 6}$

$= \dfrac{5\sqrt{36 \times 2}}{54}$

$= \dfrac{5 \times 6\sqrt{2}}{54}$

$= \dfrac{5\sqrt{2}}{9}$

3 **a** $\dfrac{4\sqrt{10}}{\sqrt{18}}$

$= \dfrac{4\sqrt{10}}{\sqrt{9 \times 2}} = \dfrac{4\sqrt{10}}{3\sqrt{2}}$

$= \dfrac{4\sqrt{10}}{3\sqrt{2}} \times \dfrac{\sqrt{2}}{\sqrt{2}}$

$= \dfrac{4\sqrt{20}}{6}$

$= \dfrac{2\sqrt{4 \times 5}}{3}$

$= \dfrac{4\sqrt{5}}{3}$

b $\dfrac{8\sqrt{20}}{\sqrt{28}}$

$= \dfrac{8\sqrt{4 \times 5}}{\sqrt{4 \times 7}}$

$= \dfrac{16\sqrt{5}}{2\sqrt{7}} = \dfrac{8\sqrt{5}}{\sqrt{7}}$

$= \dfrac{8\sqrt{5}}{\sqrt{7}} \times \dfrac{\sqrt{7}}{\sqrt{7}}$

$= \dfrac{8\sqrt{35}}{7}$

c $\dfrac{2\sqrt{45}}{5\sqrt{12}}$

$= \dfrac{2\sqrt{9 \times 5}}{5\sqrt{4 \times 3}} = \dfrac{6\sqrt{5}}{10\sqrt{3}} = \dfrac{3\sqrt{5}}{5\sqrt{3}}$

$= \dfrac{3\sqrt{5}}{5\sqrt{3}} \times \dfrac{\sqrt{3}}{\sqrt{3}}$

$= \dfrac{3\sqrt{15}}{15}$

$= \dfrac{\sqrt{15}}{5}$

d $\dfrac{5\sqrt{50}}{3\sqrt{75}}$

$= \dfrac{5\sqrt{25 \times 2}}{3\sqrt{25 \times 3}}$

$= \dfrac{25\sqrt{2}}{15\sqrt{3}} = \dfrac{5\sqrt{2}}{3\sqrt{3}}$

$= \dfrac{5\sqrt{2}}{3\sqrt{3}} \times \dfrac{\sqrt{3}}{\sqrt{3}}$

$= \dfrac{5\sqrt{6}}{9}$

e $\dfrac{6\sqrt{44}}{2\sqrt{20}}$

$= \dfrac{3\sqrt{4 \times 11}}{\sqrt{4 \times 5}} = \dfrac{6\sqrt{11}}{2\sqrt{5}} = \dfrac{3\sqrt{11}}{\sqrt{5}}$

$= \dfrac{3\sqrt{11}}{\sqrt{5}} \times \dfrac{\sqrt{5}}{\sqrt{5}}$

$= \dfrac{3\sqrt{55}}{5}$

f $\dfrac{5\sqrt{63}}{8\sqrt{32}}$

$= \dfrac{5\sqrt{9 \times 7}}{8\sqrt{16 \times 2}} = \dfrac{15\sqrt{7}}{32\sqrt{2}}$

$= \dfrac{15\sqrt{7}}{32\sqrt{2}} \times \dfrac{\sqrt{2}}{\sqrt{2}}$

$= \dfrac{15\sqrt{14}}{64}$

3 **g** $\dfrac{3\sqrt{128}}{4\sqrt{45}}$

$= \dfrac{3\sqrt{64 \times 2}}{4\sqrt{9 \times 5}}$

$= \dfrac{24\sqrt{2}}{12\sqrt{5}} = \dfrac{2\sqrt{2}}{\sqrt{5}}$

$= \dfrac{2\sqrt{2}}{\sqrt{5}} \times \dfrac{\sqrt{5}}{\sqrt{5}}$

$= \dfrac{2\sqrt{10}}{5}$

h $\dfrac{3\sqrt{125}}{5\sqrt{48}}$

$= \dfrac{3\sqrt{25 \times 5}}{5\sqrt{16 \times 3}}$

$= \dfrac{15\sqrt{5}}{20\sqrt{3}} = \dfrac{3\sqrt{5}}{4\sqrt{3}}$

$= \dfrac{3\sqrt{5}}{4\sqrt{3}} \times \dfrac{\sqrt{3}}{\sqrt{3}}$

$= \dfrac{3\sqrt{15}}{12}$

$= \dfrac{\sqrt{15}}{4}$

i $\dfrac{2\sqrt{98}}{7\sqrt{27}}$

$= \dfrac{2\sqrt{49 \times 2}}{7\sqrt{9 \times 3}} = \dfrac{14\sqrt{2}}{21\sqrt{3}} = \dfrac{2\sqrt{2}}{3\sqrt{3}}$

$= \dfrac{2\sqrt{2}}{3\sqrt{3}} \times \dfrac{\sqrt{3}}{\sqrt{3}}$

$= \dfrac{2\sqrt{6}}{9}$

$= \dfrac{2\sqrt{6}}{9}$

Challenge Yourself

1 **a** $\frac{3\sqrt{3}}{2} \times \frac{5}{\sqrt{6}}$

$= \frac{15\sqrt{3}}{2\sqrt{6}}$

$= \frac{15\sqrt{3}}{2\sqrt{6}} \times \frac{\sqrt{6}}{\sqrt{6}} = \frac{15\sqrt{18}}{12}$

$= \frac{5\sqrt{9 \times 2}}{4}$

$= \frac{15\sqrt{2}}{4}$

b $\frac{5\sqrt{18}}{3} \times \frac{\sqrt{5}}{6\sqrt{3}}$

$= \frac{5\sqrt{9 \times 2}}{3} \times \frac{\sqrt{5}}{6\sqrt{3}}$

$= \frac{15\sqrt{2}}{3} \times \frac{\sqrt{5}}{6\sqrt{3}}$

$= \frac{15\sqrt{10}}{18\sqrt{3}} = \frac{5\sqrt{10}}{6\sqrt{3}}$

$= \frac{5\sqrt{10}}{6\sqrt{3}} \times \frac{\sqrt{3}}{\sqrt{3}}$

$= \frac{5\sqrt{30}}{18}$

c $\frac{\sqrt{20}}{5\sqrt{2}} \times \frac{2}{\sqrt{27}}$

$= \frac{\sqrt{4 \times 5}}{5\sqrt{2}} \times \frac{2}{\sqrt{9 \times 3}}$

$= \frac{2\sqrt{5}}{5\sqrt{2}} \times \frac{2}{3\sqrt{3}} = \frac{4\sqrt{5}}{15\sqrt{6}}$

$= \frac{4\sqrt{5}}{15\sqrt{6}} \times \frac{\sqrt{6}}{\sqrt{6}}$

$= \frac{4\sqrt{30}}{15 \times 6}$

$= \frac{2\sqrt{30}}{15 \times 3}$

$= \frac{2\sqrt{30}}{45}$

d $\frac{\sqrt{12}}{3\sqrt{5}} \times \frac{4\sqrt{15}}{\sqrt{2}}$

$= \frac{\sqrt{4 \times 3}}{3\sqrt{5}} \times \frac{4\sqrt{15}}{\sqrt{2}}$

$= \frac{2\sqrt{3}}{3\sqrt{5}} \times \frac{4\sqrt{15}}{\sqrt{2}}$

$= \frac{8\sqrt{45}}{3\sqrt{10}} = \frac{8\sqrt{9 \times 5}}{3\sqrt{10}}$

$= \frac{24\sqrt{5}}{3\sqrt{10}} = \frac{8\sqrt{5}}{\sqrt{10}}$

$= \frac{8\sqrt{5}}{\sqrt{10}} \times \frac{\sqrt{10}}{\sqrt{10}}$

$= \frac{8\sqrt{50}}{10}$

$= \frac{8\sqrt{25 \times 2}}{10} = \frac{40\sqrt{2}}{10}$

$= 4\sqrt{2}$

2 **a**

$$\frac{3\sqrt{28}}{\sqrt{18}} \div \frac{9\sqrt{12}}{2\sqrt{4}}$$

$$= \frac{3\sqrt{28}}{\sqrt{18}} \times \frac{2\sqrt{4}}{9\sqrt{12}}$$

$$= \frac{3\sqrt{4 \times 7}}{\sqrt{9 \times 2}} \times \frac{4}{9\sqrt{4 \times 3}}$$

$$= \frac{6\sqrt{7}}{3\sqrt{2}} \times \frac{4}{18\sqrt{3}}$$

$$= \frac{\sqrt{7}}{\sqrt{2}} \times \frac{4}{9\sqrt{3}} = \frac{4\sqrt{7}}{9\sqrt{6}}$$

$$= \frac{4\sqrt{7}}{9\sqrt{6}} \times \frac{\sqrt{6}}{\sqrt{6}}$$

$$= \frac{4\sqrt{42}}{54}$$

$$= \frac{2\sqrt{42}}{27}$$

b

$$\frac{4\sqrt{3}}{\sqrt{72}} \div \frac{3\sqrt{5}}{2\sqrt{12}}$$

$$= \frac{4\sqrt{3}}{\sqrt{72}} \times \frac{2\sqrt{12}}{3\sqrt{5}}$$

$$= \frac{4\sqrt{3}}{\sqrt{36 \times 2}} \times \frac{2\sqrt{4 \times 3}}{3\sqrt{5}}$$

$$= \frac{4\sqrt{3}}{6\sqrt{2}} \times \frac{4\sqrt{3}}{3\sqrt{5}}$$

$$= \frac{4 \times 4 \times 3}{18\sqrt{10}} = \frac{48}{18\sqrt{10}}$$

$$= \frac{8}{3\sqrt{10}} \times \frac{\sqrt{10}}{\sqrt{10}}$$

$$= \frac{8\sqrt{10}}{30}$$

$$= \frac{4\sqrt{10}}{15}$$

3 **a**

$$\frac{\sqrt{125} - \sqrt{80}}{\sqrt{98}}$$

$$= \frac{\sqrt{25 \times 5} - \sqrt{16 \times 5}}{\sqrt{49 \times 2}}$$

$$= \frac{5\sqrt{5} - 4\sqrt{5}}{7\sqrt{2}}$$

$$= \frac{\sqrt{5}}{7\sqrt{2}}$$

$$= \frac{\sqrt{5}}{7\sqrt{2}} \times \frac{\sqrt{2}}{\sqrt{2}}$$

$$= \frac{\sqrt{10}}{14}$$

b

$$\frac{\sqrt{75} - \sqrt{27}}{\sqrt{90}}$$

$$= \frac{\sqrt{25 \times 3} - \sqrt{9 \times 3}}{\sqrt{9 \times 10}}$$

$$= \frac{5\sqrt{3} - 3\sqrt{3}}{3\sqrt{10}} = \frac{2\sqrt{3}}{3\sqrt{10}}$$

$$= \frac{2\sqrt{3}}{3\sqrt{10}} \times \frac{\sqrt{10}}{\sqrt{10}}$$

$$= \frac{2\sqrt{30}}{30}$$

$$= \frac{\sqrt{30}}{15}$$

Linear
Equations

Basic Equations

Multiple Choice

1 **Option D**

$7x = 28$

$x = \frac{28}{7}$

$= 4$

2 **Option C**

$\frac{5x}{3} = 10$

$5x = 10 \times 3 = 30$

$x = \frac{30}{5}$

$x = 6$

3 **Option B**

$3x + 5 = 11$

$3x = 11 - 5 = 6$

$x = \frac{6}{3}$

$x = 2$

4 **Option D**

$\frac{x}{6} - 1 = 2$

$\frac{x}{6} = 2 + 1 = 3$

$x = 3 \times 6$

$x = 18$

5 **Option A**

$\frac{3x}{4} + \frac{1}{2} = \frac{7}{2}$

$\frac{3x}{4} = \frac{7}{2} - \frac{1}{2} = \frac{6}{2} = 3$

$3x = 3 \times 4 = 12$

$x = \frac{12}{3}$

$x = 4$

Time To Practice

1

a

$x + 6 = 8$

$x = 8 - 6$

$x = 2$

b

$x - 7 = 4$

$x = 4 + 7$

$x = 11$

c

$8 - x = 3$

$-x = 3 - 8$

$-x = -5$

$x = 5$

d

$6a = -42$

$a = \frac{-42}{6}$

$a = -7$

e

$9a = 63$

$a = \frac{63}{9}$

$a = 7$

f

$4a = -44$

$a = \frac{-44}{4}$

$a = -11$

g

$\frac{m}{4} = 3$

$m = 3 \times 4$

$m = 12$

h

$\frac{3m}{5} = 12$

$3m = 12 \times 5$

$3m = 60$

$m = \frac{60}{3}$

$m = 20$

i

$\frac{7m}{10} = -7$

$7m = -7 \times 10$

$7m = -70$

$m = \frac{-70}{7}$

$m = -10$

2

a

$2x + 9 = 15$

$2x = 15 - 9$

$2x = 6$

$x = \frac{6}{2}$

$x = 3$

b

$4x + 3 = 23$

$4x = 23 - 3$

$4x = 20$

$x = \frac{20}{4}$

$x = 5$

c

$3x + 5 = -4$

$3x = -4 - 5$

$3x = -9$

$x = \frac{-9}{3}$

$x = -3$

2 **d** $6b - 12 = 48$

$6b = 48 + 12$
$6b = 60$

$b = \frac{60}{6}$

$b = 10$

e $-3b + 5 = 17$

$-3b = 17 - 5$
$-3b = 12$

$b = \frac{12}{-3}$

$b = -4$

f $-2b + 6 = 4$

$-2b = 4 - 6$
$-2b = -2$

$b = \frac{-9}{-2}$

$b = 1$

g $-4k - 9 = 3$

$-4k = 3 + 9$
$-4k = 12$

$-4k = \frac{12}{-4}$

$k = -3$

h $-3k - 6 = -3$

$-3k = -3 + 6$
$-3k = 3$

$k = \frac{3}{-3}$

$k = -1$

i $6 - 5k = 16$

$-5k = 16 - 6$
$-5k = 10$

$k = \frac{10}{-5}$

$x = -2$

3 **a** $\frac{x}{2} = 10$

$x = 10 \times 2$

$x = 20$

b $\frac{4x}{7} = -4$

$4x = -4 \times 7$

$4x = -28$

$x = \frac{-28}{4}$

$x = -7$

c $-\frac{2x}{3} = -6$

$-2x = -6 \times 3$

$-2x = -18$

$x = \frac{-18}{-2}$

$x = 9$

d $\frac{y}{2} - 5 = 3$

$\frac{y}{2} = 3 + 5$

$\frac{y}{2} = 8$

$y = 8 \times 2$

$y = 16$

e $\frac{3y}{2} + 2 = 8$

$\frac{3y}{2} = 8 - 2 = 6$

$3y = 6 \times 2 = 12$

$y = \frac{12}{3}$

$y = 4$

f $\frac{3y}{5} - 1 = 8$

$\frac{3y}{5} = 8 + 1 = 9$

$3y = 9 \times 5 = 45$

$y = \frac{45}{3}$

$y = 15$

3 **g** $-\frac{4g}{7} - 1 = 11$

$-\frac{4g}{7} = 11 + 1 = 12$

$-4g = 12 \times 7 = 84$

$g = \frac{84}{-4}$

$g = -21$

h $6 - \frac{4x}{5} = -2$

$-\frac{4x}{5} = -2 - 6 = -8$

$-4x = -8 \times 5 = -40$

$x = \frac{-40}{-4}$

$x = 10$

i $10 - \frac{7x}{9} = 3$

$-\frac{7x}{9} = 3 - 10 = -7$

$-7x = -7 \times 9 = -63$

$x = \frac{-63}{-7}$

$g = 9$

j $\frac{3x}{2} + 2 = 8$

$\frac{3x}{2} = 8 - 2 = 6$

$3x = 6 \times 2 = 12$

$x = \frac{12}{3}$

$x = 4$

k $5 - \frac{y}{2} = 1$

$-\frac{y}{2} = 1 - 5 = -4$

$-y = -4 \times 2 = -8$

$y = \frac{-8}{-1}$

$y = 8$

l $4 - \frac{2z}{3} = 0$

$\frac{2z}{3} = 0 - 4 = -4$

$2z = -4 \times 3 = -12$

$z = \frac{-12}{2}$

$x = 6$

Challenge Yourself

1 a

$$2a - \frac{3}{5} = 1$$

$$2a = 1 + \frac{3}{5}$$

$$2a = \frac{5}{5} + \frac{3}{5} = \frac{8}{5}$$

$$a = \frac{8}{5} \div 2$$

$$a = \frac{8}{5} \times \frac{1}{2}$$

$$a = \frac{8}{10}$$

$$a = \frac{4}{5}$$

b

$$\frac{a}{3} - 2 = \frac{1}{4}$$

$$\frac{a}{3} = \frac{1}{4} + 2$$

$$\frac{a}{3} = \frac{1}{4} + \frac{8}{4} = \frac{9}{4}$$

$$\frac{a}{3} = \frac{9}{4}$$

$$a = \frac{9}{4} \times 3$$

$$a = \frac{27}{4}$$

c

$$\frac{5k}{6} + 4 = -\frac{1}{3}$$

$$\frac{5k}{6} = -\frac{1}{3} - 4 = -\frac{1}{3} - \frac{12}{3} = -\frac{13}{3}$$

$$\frac{5k}{6} = -\frac{13}{3}$$

$$5k = -\frac{13}{3} \times 6$$

$$5k = -26$$

$$k = -\frac{26}{5}$$

d

$$\frac{7k}{9} - 1 = \frac{2}{3}$$

$$\frac{7k}{9} = \frac{2}{3} + 1 = \frac{2}{3} + \frac{3}{3} = \frac{5}{3}$$

$$\frac{7k}{9} = \frac{5}{3}$$

$$7k = \frac{5}{3} \times 9 = 15$$

$$k = \frac{15}{7}$$

1 **e**

$$-\frac{3p}{8}+\frac{1}{2}=\frac{5}{6}$$

$$-\frac{3p}{8}=\frac{5}{6}-\frac{1}{2}$$

$$-\frac{3p}{8}=\frac{5}{6}-\frac{3}{6}=\frac{2}{6}=\frac{1}{3}$$

$$-\frac{3p}{8}=\frac{1}{3}$$

$$-3p=\frac{1}{3}\times 8=\frac{8}{3}$$

$$p=\frac{8}{3}\div -3$$

$$p=\frac{8}{3}\times\frac{1}{-3}$$

$$p=-\frac{8}{9}$$

f

$$-\frac{2p}{3}-\frac{1}{9}=-\frac{3}{2}$$

$$-\frac{2p}{3}=-\frac{3}{2}+\frac{1}{9}$$

$$-\frac{2p}{3}=\frac{-27+2}{18}=-\frac{25}{18}$$

$$-\frac{2p}{3}=-\frac{25}{18}$$

$$-2p=-\frac{25}{18}\times 3=-\frac{25}{6}$$

$$p=-\frac{25}{6}\div -2$$

$$p=-\frac{25}{6}\times\frac{1}{-2}$$

$$p=\frac{25}{12}$$

2

$$3x-p=q$$

$$3x=q+p$$

$$x=\frac{q+p}{3}$$

3

$$y=mx+c$$

$$mx=y-c$$

$$x=\frac{y-c}{m}$$

4

$$ab-c=d$$

$$ab=d+c$$

$$a=\frac{d+c}{b}.$$

Variables on Both Sides

Multiple Choice

1 **Option A**

$3x + 5 = 2x + 12$

$3x - 2x = 12 - 5$

$x = 7$

2 **Option B**

$9y - 7 = 6y + 11$

$9y - 6y = 11 + 7$

$3y = 18$

$y = 6$

3 **Option B**

$7z - 3 = 5z + 13$

$7z - 5z = 13 + 3$

$2z = 16$

$z = 8$

4 **Option C**

$4a + 8 = 2a + 18$

$4a - 2a = 18 - 8$

$2a = 10$

$a = 5$

5 **Option D**

$8b + 3 = 4b + 15$

$8b - 4b = 15 - 3$

$4b = 12$

$b = 3$

Time To Practice

1 a $6 - x = 5 - 2x$

$-x + 2x = 5 - 6$

$x = -1$

b $8 - 2x = 10 - 3x$

$-2x + 3x = 10 - 8$

$x = 2$

c $8x + 45 = 6x + 15$

$8x - 6x = 15 - 45$

$2x = -30$

$x = -15$

d $7a + 20 = 10a - 10$

$20 + 10 = 10a - 7a$

$30 = 3a$

$a = 10$

e $3 - g = 17 - 3g$

$-g + 3g = 17 - 3$

$2g = 14$

$g = 7$

f $w + 8 = 14 - 5w$

$w + 5w = 14 - 8$

$6w = 6$

$w = 1$

g $15 - p = 27 - 3p$

$-p + 3p = 27 - 15$

$2p = 12$

$p = 6$

h $14g = 2g + 60$

$14g - 2g = 60$

$12g = 60$

$g = 5$

i $3h - 1 = 23 - h$

$3h + h = 23 + 1$

$4h = 24$

$h = 6$

2 a $6w + 27 = 11w + 17$

$27 - 17 = 11w - 6w$

$10 = 5w$

$w = 2$

b $5t + 4 = 10t - 11$

$4 + 11 = 10t - 5t$

$15 = 5t$

$t = 3$

2 c

$12r - 20 = 15r - 38$

$-20 + 38 = 15r - 12r$

$18 = 3r$

$r = 6$

d

$20f - 4 = 7f + 35$

$20f - 7f = 35 + 4$

$13f = 39$

$f = 3$

e

$-4q - 30 = 10q + 110$

$-30 - 110 -= 10q + 4q$

$-140 = 14q$

$q = -10$

f

$90 - 10v = 2 - 21v$

$-10v + 21v = 2 - 90$

$11v = -88$

$v = -8$

g

$17 - 3n = 52 - 10n$

$-3n + 10n = 52 - 17$

$7n = 35$

$n = 5$

h

$-5m - 9 = 27 - 11m$

$-5m + 11m = 27 + 9$

$6m = 36$

$m = 6$

i

$2 - 6x = 10 - 4x$

$2 - 10 = -4x + 6x$

$-8 = 2x$

$x = -4$

j

$-11t - 4 = 60 - 3t$

$-4 - 60 = -3t + 11t$

$-64 = 8t$

$t = -8$

Challenge Yourself

1 a

$3x + 4 = 2x + \frac{5}{2}$

$3x - 2x = \frac{5}{2} - 4$

$x = \frac{5}{2} - \frac{8}{2}$

$x = -\frac{3}{2}$

b

$5x - \frac{3}{4} = 2x + \frac{1}{2}$

$5x - 2x = \frac{1}{2} + \frac{3}{4}$

$3x = \frac{2}{4} + \frac{3}{4} = \frac{5}{4}$

$x = \frac{5}{4} \times \frac{1}{3}$

$x = \frac{5}{12}$

1 c

$$\frac{7q}{3}+1=\frac{5q}{3}+\frac{4}{5}$$

$$\frac{7q}{3}-\frac{5q}{3}=\frac{4}{5}-1$$

$$\frac{2q}{3}=\frac{4}{5}-\frac{5}{5}=-\frac{1}{5}$$

$$\frac{2q}{3}=-\frac{1}{5}$$

$$2q=-\frac{1}{5}\times 3=-\frac{3}{5}$$

$$q=-\frac{3}{5}\times\frac{1}{2}$$

$$q=-\frac{3}{10}$$

d

$$\frac{8p}{5}-\frac{2}{3}=\frac{4p}{5}-\frac{1}{6}$$

$$\frac{8p}{5}-\frac{4p}{5}=-\frac{1}{6}+\frac{2}{3}$$

$$\frac{4p}{5}=-\frac{1}{6}+\frac{4}{6}=\frac{3}{6}=\frac{1}{2}$$

$$\frac{4p}{5}=\frac{1}{2}$$

$$4p=\frac{1}{2}\times 5=\frac{5}{2}$$

$$p=\frac{5}{2}\times\frac{1}{4}$$

$$p=\frac{5}{8}$$

e

$$-\frac{s}{3}-\frac{1}{5}=-\frac{2s}{3}+\frac{3}{10}$$

$$-\frac{s}{3}+\frac{2s}{3}=\frac{3}{10}+\frac{1}{5}$$

$$\frac{s}{3}=\frac{3}{10}+\frac{2}{10}=\frac{5}{10}=\frac{1}{2}$$

$$\frac{s}{3}=\frac{1}{2}$$

$$s=\frac{1}{2}\times 3$$

$$s=\frac{3}{2}$$

f

$$\frac{7t}{4}-\frac{1}{2}=\frac{5t}{4}+\frac{1}{3}$$

$$\frac{7t}{4}-\frac{5t}{4}=\frac{1}{3}+\frac{1}{2}$$

$$\frac{2t}{4}=\frac{2+3}{6}=\frac{5}{6}$$

$$\frac{2t}{4}=\frac{5}{6}$$

$$\frac{t}{2}=\frac{5}{6}$$

$$t=\frac{5}{6}\times 2$$

$$t=\frac{5}{3}$$

Equations With Brackets

Multiple Choice

1 **Option B**

$2(a + 3) = 10$

$a + 3 = 5$

$a = 2$

2 **Option C**

$4(3x - 2) = 28$

$3x - 2 = 7$

$3x = 9$

$x = 3$

3 **Option C**

$2(y + 5) = 3y - 1$

$2y + 10 = 3y - 1$

$10 + 1 = 3y - 2y$

$y = 11$

4 **Option D**

$2(z - 3) + 3(z + 2) = 5$

$2z - 6 + 3z + 6 = 5$

$5z = 5$

$z = 1$

5 **Option A**

$5(3b + 2) = 2(7b + 3)$

$15b + 10 = 14b + 6$

$15b - 14b = 6 - 10$

$b = -4$

Time To Practice

1 a $4(x+3)=16$

$x+3=4$

$x=1$

b $2(x-3)=12$

$x-3=6$

$x=9$

c $2(x-4)=18$

$x-4=9$

$x=13$

d $3(1-2p)=27$

$1-2p=9$

$-2p=8$

$p=-4$

e $4(2q-5)=-12$

$2q-5=-3$

$2q=2$

$q=1$

f $2(5+4r)=-30$

$5+4r=-15$

$4r=-20$

$4=-5$

g $3(m+5)=2m$

$3m+15=2m$

$3m-2m=-15$

$m=-15$

h $8(2-n)=-6n$

$16-8n=-6n$

$16=-6n+8n$

$16=2n$

$n=8$

i $5(7-2s)=-15s$

$35-10s=-15s$

$-10s+15s=-35$

$5s=-35$

$s=-7$

2 **a** $2(k-3)=k-5$

$2k-6=k-5$

$2k-k=-5+6$

$k=1$

b $7(2-x)=8-x$

$14-7x=8-x$

$14-8=-x+7x$

$6=6x$

$x=1$

c $10-3c=2(4-c)$

$10-3c=8-2c$

$10-8=-2c+3c$

$c=2$

d $4r+7=3(2r-5)$

$4r+7=6r-15$

$7+15=6r-4r$

$22=2r$

$r=11$

e $2(3x-2)-3(x+1)=5$

$6x-4-3x-3=5$

$3x-7=5$

$3x=12$

$x=4$

f $3(2x+3)+2(x+4)=25$

$6x+9+2x+8=25$

$8x+17=25$

$8x=8$

$x=1$

g $2(2y-3)+3(4y-1)=23$

$4y-6+12y-3=23$

$16y-9=23$

$16y=32$

$y=2$

h $2(9-w)-3(w+16)=0$

$18-2w-3w-48=0$

$-30-5w=0$

$-5w=30$

$w=-6$

i $7(j+1)-2(2j-3)=10$

$7j+7-4j+6=10$

$3j+13=10$

$3j=-3$

$j=-1$

j $7(2s-1)-4(s+3)=-49$

$14s-7-4s-12=-49$

$10s-19=-49$

$10s=-30$

$s=-3$

3 a $5(u+3)=3(u+9)$

$5u+15=3u+27$

$5u-3u=27-15$

$2u=12$

$u=6$

b $8(v-1)=4(v+3)$

$8v-8=4v+12$

$8v-4v=12+8$

$4v=20$

$v=5$

c $9(2x-5)=3(4x+7)$

$18x-45=12x+21$

$18x-12x=21+45$

$6x=66$

$x=11$

d $5(2y+9)=3(3y+13)$

$10y+45=9y+39$

$10y-9y=39-45$

$y=-6$

e $5(l-3)=3(2l-1)$

$5l-15=6l-3$

$-15+3=6l-5l$

$l=-12$

f $2(4k+1)=5(3-k)$

$8k+2=15-5k$

$8k+5k=15-2$

$13k=13$

$k=1$

g $4(2d+5)=3(d+15)$

$8d+20=3d+45$

$8d-3d=45-20$

$5d=25$

$d=5$

h $3(4x-1)=7(2x-7)$

$12x-3=14x-49$

$-3+49=14x-12x$

$46=2x$

$x=23$

4 a $4(2v-3)=3(3v+2)-5$

$8v-12=9v+6-5$

$8v-12=9v+1$

$-12-1=9v-8v$

$v=-13$

b $4(j+3)+2=2(3j+5)-8$

$4j+12+2=6j+10-8$

$4j+14=6j+2$

$14-2=6j-4j$

$2j=12$

$j=6$

4 c $5(2w+3)-4=3(4w+5)-2$

$10w+15-4=12w+15-2$

$10w+11=12w+1$

$11-13=12w-10w$

$-2=2w$

$w=-1$

d $2(x-4)+7x=3(x+2)+4x$

$2x-8+7x=3x+6+4x$

$9x-8=7x+6$

$9x-7x=6+8$

$2x=14$

$x=7$

e $3(2m+1)-m=2(m+3)+2m$

$6m+3-m=2m+6+2m$

$5m+3=4m+6$

$5m-4m=6-3$

$m=3$

f $8n-3(2n-4)=4(n+2)+2n$

$8n-6n+12=4n+8+2n$

$2n+12=6n+8$

$12-8=6n-2n$

$4n=4$

$n=1$

Challenge Yourself

1 a $3\left(\frac{3y}{5}-1\right)=15$

$\frac{3y}{5}-1=5$

$\frac{3y}{5}=6$

$3y=30$

$y=10$

b $11\left(8-\frac{z}{4}\right)=99$

$8-\frac{z}{4}=9$

$-\frac{z}{4}=9-8$

$-\frac{z}{4}=1$

$z=-4$

1 c $\frac{2}{3}(3k-6)-3(k+1)=11+k$

$\frac{6k}{3}-\frac{12}{3}-3k-3=11+k$

$2k-4-3k-3=11+k$

$-k-7=11+k$

$-7-11=k+k$

$-18=2k$

$k=-9$

d $\frac{1}{2}(6z-8)+3(z+2)=4(2z-1)$

$\frac{6z}{2}-\frac{8}{2}+3z+6=8z-4$

$3z-4+3z+6=8z-4$

$6z+2=8z-4$

$2+4=8z-6z$

$6=2z$

$z=3$

e $4-2(x-6)=\frac{2x}{3}$

$4-2x+12=\frac{2x}{3}$

$16-2x=\frac{2x}{3}$

$16=\frac{2x}{3}+2x$

$16=\frac{2x}{3}+\frac{6x}{3}$

$16=\frac{8x}{3}$

$8x=16\times 3$

$8x=48$

$x=6$

1 f

$$\frac{7x}{5} - \frac{4x}{10} = 2\left(x + \frac{9}{2}\right)$$

$$\frac{7x}{5} - \frac{2x}{5} = 2x + 9$$

$$\frac{5x}{5} = 2x + 9$$

$$x = 2x + 9$$

$$-9 = 2x - x$$

$$x = -9$$

g

$$2\big(h + 5(h - 1)\big) = 110$$

$$(h + 5h - 5) = 55$$

$$6h - 5 = 55$$

$$6h = 60$$

$$h = 10$$

h

$$-6\big(2t - 3(t - 4)\big) = 0$$

$$(2t - 3t + 12) = 0$$

$$(-t + 12) = 0$$

$$-t + 12 = 0$$

$$t = 12$$

i

$$2\big(2(p - 3) + 3(1 - p)\big) = 0$$

$$(2p - 6 + 3 - 3p) = 0$$

$$-p - 3 = 0$$

$$-p = 3$$

$$p = -3$$

j

$$6 - 3\big(2 - 3(y + 1)\big) = 18$$

$$6 - 3(2 - 3y - 3) = 18$$

$$6 - 3(-1 - 3y) = 18$$

$$6 + 3 + 9y = 18$$

$$9 + 9y = 18$$

$$9y = 9$$

$$y = 1$$

Equations With Fractions

Multiple Choice

1 **Option C**

$$\frac{x}{2}+\frac{3x}{2}=16$$

$$\frac{4x}{2}=16$$

$$2x=16$$

$$x=8$$

2 **Option D**

$$\frac{4x}{3}+\frac{x}{3}=10$$

$$\frac{5x}{3}=10$$

$$5x=30$$

$$x=6$$

3 **Option B**

$$\frac{2y+1}{3}=y-2$$

$$2y+1=3(y-2)$$

$$2y+1=3y-6$$

$$1+6=3y-2y$$

$$y=7$$

4 **Option D**

$$\frac{2a+1}{3}=\frac{a+4}{2}$$

$$2(2a+1)=3(a+4)$$

$$4a+2=3a+12$$

$$4a-3a=12-2$$

$$a=10$$

5 **Option C**

$$\frac{3b+4}{4}=\frac{3b}{2}-2$$

$$\frac{3b+4}{4}=\frac{3b}{2}-\frac{2}{1}$$

$$\frac{3b+4}{4}=\frac{3b-4}{2}$$

$$2(3b+4)=4(3b-4)$$

$$6b+8=12b-16$$

$$8+16=12b-6b$$

$$24=6b$$

$$b=4$$

Time To Practice

1 a $\frac{x+1}{2} = 3$

$x + 1 = 6$

$x = 5$

b $\frac{y-3}{3} = 4$

$y - 3 = 12$

$y = 15$

c $\frac{z-9}{3} = -1$

$z - 9 = -3$

$z = 6$

d $\frac{3x+5}{2} = 7$

$3x + 5 = 14$

$3x = 9$

$x = 3$

e $\frac{5y-12}{3} = 11$

$5y - 12 = 33$

$5y = 45$

$y = 9$

f $\frac{3z-4}{7} = -4$

$3z - 4 = -28$

$3z = -24$

$z = -8$

g $\frac{7x-12}{3} = -25$

$7x - 12 = -75$

$7x = -63$

$x = -9$

h $\frac{29-2y}{3} = 5$

$29 - 2y = 15$

$29 - 15 = 2y$

$14 = 2y$

$y = 7$

1 i

$$\frac{100-5z}{3}=30$$

$$100-5z=90$$

$$100-90=5z$$

$$10=5z$$

$$z=2$$

j

$$\frac{8x+4}{5}=2x$$

$$8x+4=10x$$

$$4=10x-8x$$

$$4=2x$$

$$x=2$$

k

$$\frac{12-4y}{6}=-y$$

$$12-4y=-6y$$

$$-4y+6y=-12$$

$$2y=-12$$

$$y=-6$$

l

$$\frac{15-z}{3}=-2z$$

$$15-z=-6z$$

$$-z+6z=-15$$

$$5z=-15$$

$$z=-3$$

2 a

$$\frac{3x-1}{4}+6=5$$

$$\frac{3x-1}{4}+\frac{6}{1}=5$$

$$\frac{(3x-1)+24}{4}=5$$

$$\frac{3x+23}{4}=5$$

$$3x+23=20$$

$$3x=-3$$

$$x=-1$$

b

$$\frac{12-3y}{3}-3=5$$

$$\frac{12-3y}{3}-\frac{3}{1}=5$$

$$\frac{(12-3y)-9}{3}=5$$

$$\frac{3-3y}{3}=5$$

$$3-3y=15$$

$$-3y=12$$

$$y=-4$$

2 **c**

$$\frac{35-2z}{7}-3=2$$

$$\frac{35-2z}{7}-\frac{3}{1}=2$$

$$\frac{(35-2z)-21}{7}=2$$

$$\frac{14-2z}{7}=2$$

$$14-2z=14$$

$$14-14=2z$$

$$0=2z$$

$$z=0$$

d

$$\frac{7x+4}{3}-8=-9$$

$$\frac{7x+4}{3}-\frac{8}{1}=-9$$

$$\frac{(7x+4)-24}{3}=-9$$

$$\frac{7x-20}{3}=-9$$

$$7x-20=-27$$

$$7x=-7$$

$$x=-1$$

e

$$\frac{3y+1}{4}-3=-5$$

$$\frac{3y+1}{4}-\frac{3}{1}=-5$$

$$\frac{(3y+1)-12}{4}=-5$$

$$\frac{3y-11}{4}=-5$$

$$3y-11=-20$$

$$3y=-9$$

$$y=-3$$

f

$$\frac{4z-6}{3}-7=3$$

$$\frac{4z-6}{3}-\frac{7}{1}=3$$

$$\frac{(4z-6)-21}{3}=3$$

$$\frac{4z-27}{3}=3$$

$$4z-27=9$$

$$4z=36$$

$$z=9$$

3 **a**

$$\frac{x}{3} - \frac{x}{2} = 1$$

$$\frac{2x - 3x}{6} = 1$$

$$\frac{-x}{6} = 1$$

$$-x = 6$$

$$x = -6$$

b

$$\frac{y}{4} - \frac{y}{5} = 1$$

$$\frac{5y - 4y}{20} = 1$$

$$\frac{y}{20} = 1$$

$$y = 20$$

c

$$\frac{5z}{4} - \frac{z}{2} = -\frac{3}{4}$$

$$\frac{5z}{4} - \frac{2z}{4} = -\frac{3}{4}$$

$$\frac{5z - 2z}{4} = -\frac{3}{4}$$

$$\frac{3z}{4} = -\frac{3}{4}$$

$$3z = -3$$

$$z = -1$$

d

$$\frac{x}{4} - \frac{x}{7} = \frac{9}{14}$$

$$\frac{7x - 4x}{28} = \frac{9}{14}$$

$$\frac{3x}{28} = \frac{9}{14}$$

$$3x = \frac{9}{14} \times 28$$

$$3x = 18$$

$$x = 6$$

e

$$y + \frac{y}{7} = 8$$

$$\frac{y}{1} + \frac{y}{7} = 8$$

$$\frac{7y + y}{7} = 8$$

$$\frac{8y}{7} = 8$$

$$8y = 56$$

$$y = 7$$

f

$$-z - \frac{z}{4} = \frac{5}{4}$$

$$\frac{-z}{1} - \frac{z}{4} = \frac{5}{4}$$

$$\frac{-4z - z}{4} = \frac{5}{4}$$

$$\frac{-5z}{4} = \frac{5}{4}$$

$$-5z = 5$$

$$z = -1$$

3 **g**

$$2x + \frac{3x}{5} = -13$$

$$\frac{2x}{1} + \frac{3x}{5} = -13$$

$$\frac{10x + 3x}{5} = -13$$

$$\frac{13x}{5} = -13$$

$$13x = -65$$

$$x = -5$$

h

$$6y - \frac{3y}{2} = 27$$

$$\frac{6y}{1} - \frac{3y}{2} = 27$$

$$\frac{12y - 3y}{2} = 27$$

$$\frac{9y}{2} = 27$$

$$9y = 54$$

$$y = 6$$

i

$$-4z + \frac{5z}{2} = -15$$

$$\frac{-4z}{1} + \frac{5z}{2} = -15$$

$$\frac{-8z + 5z}{2} = -15$$

$$\frac{-3z}{2} = -15$$

$$-3z = -30$$

$$z = 10$$

4 **a**

$$\frac{2x + 1}{3} = x - 2$$

$$2x + 1 = 3(x - 2)$$

$$2x + 1 = 3x - 6$$

$$1 + 6 = 3x - 2x$$

$$x = 7$$

b

$$\frac{5y - 3}{2} = 2y + 4$$

$$5y - 3 = 2(2y + 4)$$

$$5y - 3 = 4y + 8$$

$$5y - 4y = 8 + 3$$

$$y = 11$$

4 **c** $\dfrac{z+17}{5}=z+1$

$z+17=5(z+1)$

$z+17=5z+5$

$17-5=5z-z$

$12=4z$

$z=3$

d $4x-9=\dfrac{2x+3}{3}$

$3(4x-9)=2x+3$

$12x-27=2x+3$

$12x-2x=3+27$

$10x=30$

$x=3$

e $y+3=\dfrac{9-y}{2}$

$2(y+3)=9-y$

$2y+6=9-y$

$2y+y=9-6$

$3y=3$

$y=1$

f $-3z-2=\dfrac{8z-33}{5}$

$5(-3z-2)=8z-33$

$-15z-10=8z-33$

$-10+33=8z+15z$

$23=23z$

$z=1$

5 **a** $\dfrac{3x+7}{5}=\dfrac{2x}{3}$

$3(3x+7)=10x$

$9x+21=10x$

$21=10x-9x$

$x=21$

b $\dfrac{3y-10}{7}=\dfrac{2y}{5}$

$5(3y-10)=14y$

$15y-50=14y$

$15y-14y=50$

$y=50$

c $\dfrac{z-8}{3}=\dfrac{z}{6}$

$6(z-8)=3z$

$6z-48=3z$

$6z-3z=48$

$3z=48$

$z=16$

d $\dfrac{x-5}{3}=\dfrac{x-7}{4}$

$4(x-5)=3(x-7)$

$4x-20=3x-21$

$4x-3x=-21+20$

$x=-1$

5 **e** $\frac{y+2}{6}=\frac{y-5}{3}$

$3(y+2)=6(y-5)$

$3y+6=6y-30$

$6+30=6y-3y$

$36=3y$

$y=12$

f $\frac{z+3}{4}=\frac{27-z}{6}$

$6(z+3)=4(27-z)$

$6z+18=108-4z$

$6z+4z=108-18$

$10z=90$

$z=9$

g $\frac{9x+3}{2}=\frac{13x+7}{3}$

$3(9x+3)=2(13x+7)$

$27x+9=26x+14$

$27x-26x=14-9$

$x=5$

h $\frac{4y+9}{15}=\frac{y+2}{5}$

$5(4y+9)=15(y+2)$

$20y+45=15y+30$

$20y-15y=30-45$

$5y=-15$

$y=-3$

i $\frac{2z+7}{5}=\frac{3-z}{4}$

$4(2z+7)=5(3-z)$

$8z+28=15-5z$

$8z+5z=15-28$

$13z=-13$

$z=-1$

j $\frac{x+11}{2}=\frac{x+14}{3}$

$3(x+11)=2(x+14)$

$3x+33=2x+28$

$3x-2x=28-33$

$x=-5$

k $\frac{3-z}{2}=\frac{6(9-z)}{4}$

$4(3-z)=12(9-z)$

$12-4z=108-12z$

$-4z+12z=108-12$

$8z=96$

$z=12$

l $\frac{3y-4}{2}=\frac{102-5y}{4}$

$4(3y-4)=2(102-5y)$

$12y-16=204-10y$

$12y+10y=204+16$

$22y=220$

$y=10$

6 **a**

$$\frac{x+3}{2} - 3 = \frac{x+4}{4}$$

$$\frac{x+3}{2} - \frac{3}{1} = \frac{x+4}{4}$$

$$\frac{(x+3)-6}{2} = \frac{x+4}{4}$$

$$\frac{x-3}{2} = \frac{x+4}{4}$$

$4(x-3) = 2(x+4)$

$4x - 12 = 2x + 8$

$4x - 2x = 8 + 12$

$2x = 20$

$x = 10$

b

$$\frac{x+3}{10} = 2 - \frac{5-x}{2}$$

$$\frac{x+3}{10} = \frac{2}{1} - \frac{5-x}{2}$$

$$\frac{x+3}{10} = \frac{4-(5-x)}{2}$$

$$\frac{x+3}{10} = \frac{x-1}{2}$$

$2(x+3) = 10(x-1)$

$2x + 6 = 10x - 10$

$6 + 10 = 10x - 2x$

$16 = 8x$

$x = 2$

c

$$\frac{x-1}{4} = \frac{x+3}{2} - 4$$

$$\frac{x-1}{4} = \frac{x+3}{2} - \frac{4}{1}$$

$$\frac{x-1}{4} = \frac{(x+3)-8}{2}$$

$$\frac{x-1}{4} = \frac{x-5}{2}$$

$2(x-1) = 4(x-5)$

$2x - 2 = 4x - 20$

$-2 + 20 = 4x - 2x$

$18 = 2x$

$x = 9$

d

$$\frac{4x+9}{15} = 1 + \frac{x-3}{5}$$

$$\frac{4x+9}{15} = \frac{5}{5} + \frac{x-3}{5}$$

$$\frac{4x+9}{15} = \frac{x+2}{5}$$

$5(4x+9) = 15(x+2)$

$20x + 45 = 15x + 30$

$20x - 15x = 30 - 45$

$5x = -15$

$x = -3$

6

e

$$\frac{6x+7}{5}=\frac{5x+1}{4}+1$$

$$\frac{6x+7}{5}=\frac{5x+1}{4}+\frac{4}{4}$$

$$\frac{6x+7}{5}=\frac{5x+5}{4}$$

$4(6x+7)=5(5x+5)$

$24x+28=25x+25$

$28-25=25x-24x$

$x=3$

f

$$\frac{2x+10}{3}=\frac{6-5x}{4}-2$$

$$\frac{2x+10}{3}=\frac{6-5x}{4}-\frac{2}{1}$$

$$\frac{2x+10}{3}=\frac{(6-5x)-8}{4}$$

$$\frac{2x+10}{3}=\frac{-2-5x}{4}$$

$4(2x+10)=3(-2-5x)$

$8x+40=-6-15x$

$8x+15x=-6-40$

$23x=-46$

$x=-2$

g

$$\frac{3x-4}{5}=\frac{5x+9}{7}-3$$

$$\frac{3x-4}{5}=\frac{5x+9}{7}-\frac{3}{1}$$

$$\frac{3x-4}{5}=\frac{(5x+9)-21}{7}$$

$$\frac{3x-4}{5}=\frac{5x-12}{7}$$

$7(3x-4)=5(5x-12)$

$21x-28=25x-60$

$-28+60=25x-21x$

$32=4x$

$x=8$

h

$$6-\frac{7-2x}{3}=\frac{1-2x}{9}$$

$$\frac{6}{1}-\frac{7-2x}{3}=\frac{1-2x}{9}$$

$$\frac{18-(7-2x)}{3}=\frac{1-2x}{9}$$

$$\frac{2x+11}{3}=\frac{1-2x}{9}$$

$9(2x+11)=3(1-2x)$

$18x+99=3-6x$

$18x+6x=3-99$

$24x=-96$

$x=-4$

7 **a**

$$\frac{5x-10}{3}-\frac{3x}{2}=\frac{2x}{3}$$

$$\frac{2(5x-10)-9x}{6}=\frac{2x}{3}$$

$$\frac{10x-20-9x}{6}=\frac{2x}{3}$$

$$\frac{x-20}{6}=\frac{2x}{3}$$

$$3(x-20)=12x$$

$$3x-60=12x$$

$$-60=12x-3x$$

$$-60=9x$$

$$x=\frac{-60}{9}=-\frac{20}{3}$$

b

$$\frac{-3x+12}{6}+\frac{2x-6}{3}=-\frac{x}{2}$$

$$\frac{-3x+12}{6}+\frac{2(2x-6)}{6}=-\frac{x}{2}$$

$$\frac{-3x+12+4x-12}{6}=-\frac{x}{2}$$

$$\frac{x}{6}=-\frac{x}{2}$$

$$2x=-6x$$

$$2x+6x=0$$

$$8x=0$$

$$x=0$$

c

$$\frac{2x-4}{3}+\frac{x+5}{6}=\frac{2-x}{4}$$

$$\frac{2(2x-4)}{6}+\frac{x+5}{6}=\frac{2-x}{4}$$

$$\frac{4x-8+x+5}{6}=\frac{2-x}{4}$$

$$\frac{5x-3}{6}=\frac{2-x}{4}$$

$$4(5x-3)=6(2-x)$$

$$20x-12=12-6x$$

$$20x+6x=12+12$$

$$26x=24$$

$$x=\frac{24}{26}=\frac{12}{13}$$

d

$$\frac{-x-6}{5}+\frac{4x+7}{10}=\frac{3x+1}{5}$$

$$\frac{2(-x-6)}{10}+\frac{4x+7}{10}=\frac{2(3x+1)}{10}$$

$$\frac{-2x-12+4x+7}{10}=\frac{6x+2}{10}$$

$$\frac{2x-5}{10}=\frac{6x+2}{10}$$

$$2x-5=6x+2$$

$$-5-2=6x-2x$$

$$-7=4x$$

$$x=-\frac{7}{4}$$

7 **e**

$$\frac{5x+8}{5}-\frac{3x+6}{10}=\frac{x+1}{2}$$

$$\frac{2(5x+8)}{10}-\frac{3x+6}{10}=\frac{x+1}{2}$$

$$\frac{10x+16}{10}-\frac{3x+6}{10}=\frac{x+1}{2}$$

$$\frac{10x+16-(3x+6)}{10}=\frac{x+1}{2}$$

$$\frac{10x+16-3x-6}{10}=\frac{x+1}{2}$$

$$\frac{7x+10}{10}=\frac{x+1}{2}$$

$$2(7x+10)=10(x+1)$$

$$14x+20=10x+10$$

$$14x-10x=10-20$$

$$4x=-10$$

$$4x=-\frac{10}{4}=-\frac{5}{2}$$

f

$$\frac{3x-8}{3}+\frac{3-x}{6}=\frac{2x-5}{4}$$

$$\frac{2(3x-8)}{6}+\frac{3-x}{6}=\frac{2x-5}{4}$$

$$\frac{6x-16}{6}+\frac{3-x}{6}=\frac{2x-5}{4}$$

$$\frac{6x-16+3-x}{6}=\frac{2x-5}{4}$$

$$\frac{5x-13}{6}=\frac{2x-5}{4}$$

$$4(5x-13)=6(2x-5)$$

$$20x-52=12x-30$$

$$20x-12x=-30+52$$

$$8x=22$$

$$x=\frac{22}{8}=\frac{11}{4}$$

Challenge Yourself

1 **a**

$$\frac{7x}{5}-\frac{4x}{10}=2\left(x+\frac{9}{2}\right)$$

$$\frac{14x}{10}-\frac{4x}{10}=2x+9$$

$$\frac{10x}{10}=2x+9$$

$$x=2x+9$$

$$-9=2x-x$$

$$x=-9$$

1 **b**

$$\frac{3}{4} - 2\left(a + \frac{1}{5}\right) = \frac{a}{3}$$

$$\frac{3}{4} - 2a - \frac{2}{5} = \frac{a}{3}$$

$$\frac{3}{4} - \frac{2}{5} = \frac{a}{3} + \frac{2a}{1}$$

$$\frac{15 - 8}{20} = \frac{a + 6a}{3}$$

$$\frac{7}{20} = \frac{7a}{3}$$

$$21 = 140a$$

$$a = \frac{21}{140} = \frac{7 \times 3}{7 \times 20}$$

$$a = \frac{3}{20}$$

c

$$\frac{41}{9} = \frac{5}{2}\left(y + \frac{2}{3}\right) - \frac{y}{3}$$

$$\frac{41}{9} = \frac{5y}{2} + \frac{5}{3} - \frac{y}{3}$$

$$\frac{41}{9} - \frac{5}{3} = \frac{5y}{2} - \frac{y}{3}$$

$$\frac{41}{9} - \frac{15}{9} = \frac{15y - 2y}{6}$$

$$\frac{26}{9} = \frac{13y}{6}$$

$$6 \times 26 = 9 \times 13y$$

$$y = \frac{6 \times 26}{9 \times 13} = \frac{6 \times 2}{9 \times 1} = \frac{2 \times 2}{3 \times 1}$$

$$y = \frac{4}{3}$$

d

$$-k - \frac{5}{4}\left(1 - \frac{8k}{3}\right) = -\frac{19}{4}$$

$$-k - \frac{5}{4}\left(1 - \frac{8k}{3}\right) = -\frac{19}{4}$$

$$-k - \frac{5}{4} + \frac{40k}{12} = -\frac{19}{4}$$

$$-k + \frac{10k}{3} = -\frac{19}{4} + \frac{5}{4}$$

$$\frac{-k}{1} + \frac{10k}{3} = -\frac{14}{4}$$

$$\frac{-3k}{3} + \frac{10k}{3} = -\frac{7}{2}$$

$$\frac{7k}{3} = -\frac{7}{2}$$

$$14k = -21$$

$$k = -\frac{21}{14} = -\frac{7 \times 3}{7 \times 2}$$

$$k = -\frac{3}{2}$$

1 **e** $\frac{1}{2}\left(1-\frac{2n}{3}\right)=\frac{n}{5}+\frac{1}{2}$

$\frac{1}{2}-\frac{n}{6}=\frac{2n+5}{10}$

$\frac{3}{6}-\frac{n}{6}=\frac{2n+5}{10}$

$\frac{3-n}{6}=\frac{2n+5}{10}$

$10(3-n)=6(2n+5)$

$30-10n=12n+30$

$30-30=12n+10n$

$0=22n$

$n=0$

f $\frac{7}{6}-\frac{4p}{3}=2\left(p+\frac{3}{2}\right)-\frac{3p}{2}$

$\frac{7}{6}-\frac{4p}{3}=2p+3-\frac{3p}{2}$

$\frac{7}{6}-\frac{8p}{6}=\frac{2p}{1}+\frac{3}{1}-\frac{3p}{2}$

$\frac{7-8p}{6}=\frac{4p}{2}+\frac{6}{2}-\frac{3p}{2}$

$\frac{7-8p}{6}=\frac{p+6}{2}$

$2(7-8p)=6(p+6)$

$14-16p=6p+36$

$14-36=6p+16p$

$-22=22p$

$p=-\frac{22}{22}$

$p=-1$

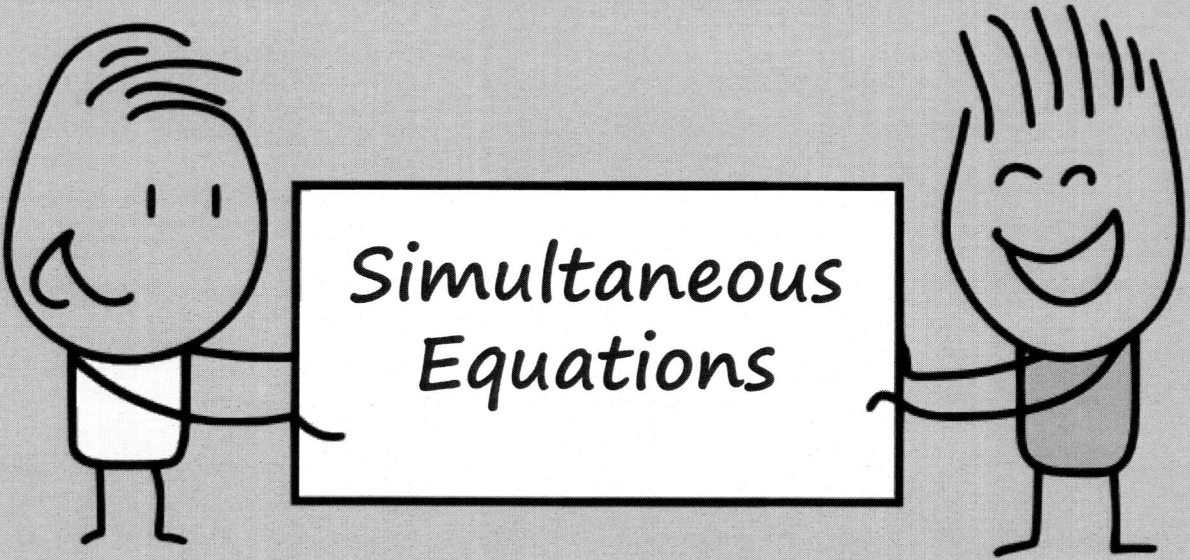
Simultaneous
Equations

The Substitution Method

Multiple Choice

1 **Option B**

$y = x + 5$ (1)
$3x + y = 13$ (2)

Sub (1) into (2)

$3x + (x + 5) = 13$
$4x + 5 = 13$
$4x = 8$
$x = 2$

2 **Option C**

$y = x + 3$ (1)
$6x + y = 17$ (2)

$x = y - 3$ (3)

Sub (3) into (2)

$6x + y = 17$
$6(y - 3) + y = 17$
$6y - 18 + y = 17$
$7y - 18 = 17$
$7y = 35$
$y = 5$

3 **Option C**

$x = y - 1$ (1)
$2x + 3y = 8$ (2)

$y = x + 1$ (3)

Sub (3) into (2)

$2x + 3y = 8$
$2x + 3(x + 1) = 8$
$2x + 3x + 3 = 8$
$5x + 3 = 8$
$5x = 5$
$x = 1$

4 **Option B**

$4a + 5b = 6$ (1)
$a - 2b = -5$ (2)

$a = 2b - 5$ (3)

Sub (3) into (1)

$4a + 5b = 6$
$4(2b - 5) + 5b = 6$
$8b - 20 + 5b = 6$
$13b - 20 = 6$
$13b = 26$
$b = 2$

Sub $b = 2$ into (3)

$a = 2b - 5$

$a = 2(2) - 5$

$a = 4 - 5$

$a = -1$

5 **Option D**

$3m + n = 5$ (1)
$5m - 4n = -3$ (2)

$n = 5 - 3m$ (3)

Sub (3) into (2)

$5m - 4n = -3$
$5m - 4(5 - 3m) = -3$

$5m - 20 + 12m = -3$

$17m - 20 = -3$

$17m = 17$
$m = 1$

Sub $m = 1$ into (3)

$n = 5 - 3m$
$n = 5 - 3(1)$
$n = 5 - 3$
$n = 2$

Time To Practice

1 **a**

$y = x$ (1)
$4x + 3y = 7$ (2)

Sub (1) into (2)

$4x + 3y = 7$
$4x + 3x = 7$
$7x = 7$
$x = 1$

Sub $x = 1$ into (1)

$y = x$
$y = 1$

$\therefore x = 1 \text{ and } y = 1$

b

$y = x$ (1)
$7x + 3y = 20$ (2)

Sub (1) into (2)

$7x + 3y = 20$
$7x + 3x = 20$
$10x = 20$
$x = 2$

Sub $x = 2$ into (1)

$y = x$
$y = 2$

$\therefore x = 2 \text{ and } y = 2$

c

$y = x - 1$ (1)
$3x + 2y = 8$ (2)

Sub (1) into (2)

$3x + 2y = 8$
$3x + 2(x - 1) = 8$
$3x + 2x - 2 = 8$
$5x - 2 = 8$
$5x = 10$
$x = 2$

Sub $x = 2$ into (1)

$y = x - 1$
$y = 2 - 1$
$y = 1$

$\therefore x = 2 \text{ and } y = 1$

d

$y = x + 1$ (1)
$3x - 2y = -5$ (2)

Sub (1) into (2)

$3x - 2y = -5$
$3x - 2(x + 1) = -5$
$3x - 2x - 2 = -5$
$x - 2 = -5$
$x = -3$

Sub $x = -3$ into (1)

$y = x + 1$
$y = -3 + 1$
$y = -2$

$\therefore x = -3 \text{ and } y = -2$

1 e

$x = 2y + 1$ (1)
$11y - 5x = -2$ (2)

Sub (1) into (2)

$11y - 5x = -2$
$11y - 5(2y + 1) = -2$
$11y - 10y - 5 = -2$
$y - 5 = -2$
$y = 3$

Sub $y = 1$ into (1)

$x = 2y + 1$
$x = 2(3) + 1$
$x = 7$

$\therefore x = 7 \; and \; y = 3$

f

$x = 3y - 3$ (1)
$7y - 2x = 11$ (2)

Sub (1) into (2)

$7y - 2x = 11$
$7y - 2(3y - 3) = 11$
$7y - 6y + 6 = 11$
$y + 6 = 11$
$y = 5$

Sub $y = 5$ into (1)

$x = 3y - 3$
$x = 3(5) - 3$
$x = 12$

$\therefore x = 12 \; and \; y = 5$

g

$x = 2y + 12$ (1)
$3x + 2y = -4$ (2)

Sub (1) into (2)

$3x + 2y = -4$
$3(2y + 12) + 2y = -4$
$6y + 36 + 2y = -4$
$8y + 36 = -4$
$8y = -40$
$y = -5$

Sub $y = -5$ into (1)

$x = 2y + 12$
$x = 2(-5) + 12$
$x = -10 + 12$
$x = 2$

$\therefore x = 2 \; and \; y = -5$

h

$x = 4y + 2$ (1)
$2y - 3x = 14$ (2)

Sub (1) into (2)

$2y - 3x = 14$
$2y - 3(4y + 2) = 14$
$2y - 12y - 6 = 14$
$-10y - 6 = 14$
$-10y = 20$
$y = -2$

Sub $y = -2$ into (1)

$x = 4y + 2$
$x = 4(-2) + 2$
$x = -8 + 2$
$x = -6$

$\therefore x = -6 \; and \; y = -2$

2 **a**

$3x + 4y = 27$ (1)
$2x - y = -4$ (2)

Rearrange (2)

$2x - y = -4$
$y = 2x + 4$ (3)

Sub (3) into (1)

$3x + 4y = 27$
$3x + 4(2x + 4) = 27$
$3x + 8x + 16 = 27$
$11x + 16 = 27$
$11x = 11$
$x = 1$

Sub $x = 1$ into (3)

$y = 2x + 4$
$y = 2(1) + 4$
$y = 6$

$\therefore x = 1 \; and \; y = 6$

b

$3x - y = 23$ (1)
$2x - 2y = 10$ (2)

Rearrange (1)

$3x - y = 23$
$y = 3x - 23$ (3)

Sub (3) into (2)

$2x - 2y = 10$
$2x - 2(3x - 23) = 10$
$2x - 6x + 46 = 10$
$-4x + 46 = 10$
$-4x = -36$
$x = 9$

Sub $x = 9$ into (3)

$y = 3x - 23$
$y = 3(9) - 23$
$y = 4$

$\therefore x = 9 \; and \; y = 4$

2 **c**

$4a + 3b = 23$ (1)
$2a - b = -1$ (2)

Rearrange (2)

$2a - b = -1$
$b = 2a + 1$ (3)

Sub (3) into (1)

$4a + 3b = 23$
$4a + 3(2a + 1) = 23$
$4a + 6a + 3 = 23$
$10a + 3 = 23$
$10a = 20$
$a = 2$

Sub $a = 2$ into (3)

$b = 2a + 1$
$b = 2(2) + 1$
$b = 5$

$\therefore a = 2 \text{ and } b = 5$

d

$2a + b = 16$ (1)
$3a - 4b = 24$ (2)

Rearrange (1)

$2a + b = 16$
$b = 16 - 2a$ (3)

Sub (3) into (2)

$3a - 4b = 24$
$3a - 4(16 - 2a) = 24$
$3a - 64 + 8a = 24$
$11a - 64 = 24$
$11a = 88$
$a = 8$

Sub $a = 8$ into (3)

$b = 16 - 2a$
$b = 16 - 2(8)$
$b = 0$

$\therefore x = 8 \text{ and } y = 0$

2 **e**

$5p - q = -22$ (1)
$7p + 2q = -7$ (2)

Rearrange (1)

$5p - q = -22$
$q = 5p + 22$ (3)

Sub (3) into (2)

$7p + 2q = -7$
$7p + 2(5p + 22) = -7$
$7p + 10p + 44 = -7$
$17p + 44 = -7$
$17p = -51$
$p = -3$

Sub $p = -3$ into (3)

$q = 5p + 22$
$q = 5(-3) + 22$
$q = -15 + 22$
$q = 7$

$\therefore p = -3 \; and \; q = 7$

f

$5p + 2q = 5$ (1)
$3p - q = 25$ (2)

Rearrange (2)

$3p - q = 25$
$q = 3p - 25$ (3)

Sub (3) into (1)

$5p + 2q = 5$
$5p + 2(3p - 25) = 5$
$5p + 6p - 50 = 5$
$11p - 50 = 5$
$11p = 55$
$p = 5$

Sub $p = 5$ into (3)

$q = 3p - 25$
$q = 3(5) - 25$
$q = 15 - 25$
$q = -10$

$\therefore p = 5 \; and \; q = -10$

2 g

$3m - 2n = 34$ (1)
$m + 4n = 16$ (2)

Rearrange (2)

$m + 4n = 16$
$m = 16 - 4n$ (3)

Sub (3) into (1)

$3m - 2n = 34$
$3(16 - 4n) - 2n = 34$
$48 - 12n - 2n = 34$
$48 - 14n = 34$
$48 - 34 = 14n$
$14 = 14n$
$n = 1$

Sub $n = 1$ into (3)

$m = 16 - 4n$
$m = 16 - 4(1)$
$m = 12$

$\therefore m = 12 \; and \; n = 1$

h

$6m - 3n = 21$ (1)
$2m + n = -15$ (2)

Rearrange (2)

$2m + n = -15$
$n = -15 - 2m$ (3)

Sub (3) into (1)

$6m - 3n = 21$
$6m - 3(-15 - 2m) = 21$
$6m + 45 + 6m = 21$
$12m + 45 = 21$
$12m = -24$
$m = -2$

Sub $m = -2$ into (3)

$n = -15 - 2m$
$n = -15 - 2(-2)$
$n = -15 + 4$
$n = -11$

$\therefore m = -2 \; and \; n = -11$

2 **i** $5r + 3s = 44$ (1)
$2r - s = 0$ (2)

Rearrange (2)

$2r - s = 0$
$s = 2r$ (3)

Sub (3) into (1)

$5r + 3s = 44$
$5r + 3(2r) = 44$
$5r + 6r = 44$
$11r = 44$
$r = 4$

Sub $r = 4$ into (3)

$s = 2r$
$s = 2(4)$
$s = 8$

$\therefore r = 4 \text{ and } s = 8$

j $4r + 3s = -59$ (1)
$r - 2s = 10$ (2)

Rearrange (2)

$r - 2s = 10$
$r = 2s + 10$ (3)

Sub (3) into (1)

$4r + 3s = -59$
$4(2s + 10) + 3s = -59$
$8s + 40 + 3s = -59$
$11s + 40 = -59$
$11s = -99$
$s = -9$

Sub $s = -9$ into (3)

$r = 2s + 10$
$r = 2(-9) + 10$
$r = -18 + 10$
$r = -8$

$\therefore r = -8 \text{ and } s = -9$

Challenge Yourself

1 a $m - 2n - 5 = 0$ (1)

$3m - n - 25 = 0$ (2)

Rearrange (1)

$m - 2n - 5 = 0$

$m = 2n + 5$ (3)

Sub (3) into (2)

$3m - n - 25 = 0$

$3(2n + 5) - n - 25 = 0$

$6n + 15 - n - 25 = 0$

$5n - 10 = 0$

$5n = 10$

$n = 2$

Sub $n = 2$ into (3)

$m = 2n + 5$

$m = 2(2) + 5$

$m = 9$

$\therefore m = 9 \text{ and } n = 2$

b $y + x - 10 = 0$ (1)

$7y + 2x - 5 = 0$ (2)

Rearrange (1)

$y + x - 10 = 0$

$x = 10 - y$ (3)

Sub (3) into (2)

$7y + 2x - 5 = 0$

$7y + 2(10 - y) - 5 = 0$

$7y + 20 - 2y - 5 = 0$

$5y + 15 = 0$

$5y = -15$

$y = -3$

Sub $y = -3$ into (3)

$x = 10 - y$

$x = 10 - (-3)$

$x = 13$

$\therefore x = 13 \text{ and } y = -3$

1 **c**

$$r - 2s = 4 \quad (1)$$
$$r + \frac{s}{3} = 4 \quad (2)$$

Rearrange (1)

$$r - 2s = 4$$
$$r = 2s + 4 \quad (3)$$

Sub (3) into (2)

$$r + \frac{s}{3} = 4$$

$$2s + 4 + \frac{s}{3} = 4$$

$$2s + \frac{s}{3} = 0$$

$$\frac{2s}{1} + \frac{s}{3} = 0$$

$$\frac{6s + 3s}{3} = 0$$

$$9s = 0$$

$$s = 0$$

Sub $s = 0$ into (3)

$$r = 2s + 4$$
$$r = 2(0) + 4$$
$$r = 4$$

$$\therefore r = 4 \text{ and } s = 0$$

d

$$2c + \frac{d}{4} = 7 \quad (1)$$
$$c - \frac{d}{2} = 1 \quad (2)$$

Rearrange (2)

$$c - \frac{d}{2} = 1$$
$$c = \frac{d}{2} + 1 \quad (3)$$

Sub (3) into (1)

$$2c + \frac{d}{4} = 7$$

$$2\left(\frac{d}{2} + 1\right) + \frac{d}{4} = 7$$

$$d + 2 + \frac{d}{4} = 7$$

$$d + \frac{d}{4} = 5$$

$$\frac{d}{1} + \frac{d}{4} = 5$$

$$\frac{4d + d}{4} = 5$$

$$5d = 20$$

$$d = 4$$

Sub $d = 4$ into (3)

$$c = \frac{d}{2} + 1$$
$$c = \frac{4}{2} + 1 = 3$$

$$\therefore c = 3 \text{ and } d = 4$$

1 **e**

$$3p + q = 5 \quad (1)$$
$$p - \frac{q}{3} = 1 \quad (2)$$

Rearrange (1)

$$3p + q = 5$$
$$q = 5 - 3p \quad (3)$$

Sub (3) into (2)

$$p - \frac{q}{3} = 1$$
$$p - \frac{5 - 3p}{3} = 1$$
$$\frac{p}{1} - \frac{5 - 3p}{3} = 1$$
$$\frac{3p - (5 - 3p)}{3} = 1$$
$$\frac{6p - 5}{3} = 1$$
$$6p - 5 = 3$$
$$6p = 8$$
$$p = \frac{8}{6} = \frac{4}{3}$$

Sub $p = \frac{4}{3}$ into (3)

$$q = 5 - 3p$$
$$q = 5 - 3\left(\frac{4}{3}\right)$$
$$q = 5 - 4$$
$$q = 1$$

$$\therefore p = \frac{4}{3} \text{ and } q = 1$$

f

$$2a - b = 3 \quad (1)$$
$$a + \frac{b}{2} = 2 \quad (2)$$

Rearrange (1)

$$2a - b = 3$$
$$b = 2a - 3 \quad (3)$$

Sub (3) into (2)

$$a + \frac{b}{2} = 2$$
$$a + \frac{2a - 3}{2} = 2$$
$$\frac{2a}{2} + \frac{2a - 3}{2} = 2$$
$$\frac{4a - 3}{2} = 2$$
$$4a - 3 = 4$$
$$4a = 7$$
$$a = \frac{7}{4}$$

Sub $a = \frac{7}{4}$ into (3)

$$b = 2a - 3$$
$$b = 2\left(\frac{7}{4}\right) - 3$$
$$b = \frac{7}{2} - \frac{3}{1} = \frac{7 - 6}{2}$$
$$b = \frac{1}{2}$$

$$\therefore c = \frac{7}{4} \text{ and } d = \frac{1}{2}$$

The Elimination Method

Multiple Choice

1 **Option A**

$4a - 3b = -10$ (1)
$2a + 3b = 4$ (2)

$(1) + (2)$

$4a + 2a = -10 + 4$
$6a = -6$

2 **Option D**

$-5x + 2y = 16$ (1)
$-5x - 3y = 1$ (2)

$(1) - (2)$

$2y - (-3y) = 16 - 1$
$5y = 15$

3 **Option A**

$3x + 4y = 2$ (1)
$5x + 2y = -6$ (2)

$2 \times (2)$

$2 \times (5x + 2y = -6)$ (3)
$10x + 4y = -12$

$3x + 4y = 2$ (1)
$10x + 4y = -12$ (3)

$(3) - (1)$

$10x - 3x = -12 - 2$
$7x = -14$
$x = -2$

4 **Option B**

$-2x + 3y = 7$ (1)
$3x - 2y = -3$ (2)

$3 \times (1)$ *and* $2 \times (2)$

$3 \times (-2x + 3y = 7)$
$2 \times (3x - 2y = -3)$

$-6x + 9y = 21$ (3)
$6x - 4y = -6$ (4)

$(4) + (3)$

$-4y + 9y = -6 + 21$
$5y = 15$
$y = 3$

5 **Option C**

$3x + 4y = 13$ (1)
$2x - 5y = 1$ (2)

$\mathbf{2} \times (\mathbf{1})$ *and* $\mathbf{3} \times (\mathbf{2})$

$2 \times (3x + 4y = 13)$
$3 \times (2x - 5y = 1)$

$6x + 8y = 26$ (3)
$6x - 15y = 3$ (4)

$(\mathbf{3}) - (\mathbf{4})$

$8y - (-15y) = 26 - 3$
$23y = 23$
$y = 1$

Sub $y = \mathbf{1}$ into (2)

$2x - 5y = 1$
$2x - 5(1) = 1$
$2x - 5 = 1$
$2x = 6$
$x = 3$

$\therefore x = 3 \text{ and } y = 1$

Time To Practice

1 **a**

$3x + 5y = 21$ (1)
$2x + 5y = 19$ (2)

$(1) - (2)$

$3x - 2x = 21 - 19$
$x = 2$

Sub $x = 2$ into (2)

$2x + 5y = 19$
$2(2) + 5y = 19$
$4 + 5y = 19$
$5y = 15$
$y = 3$

$\therefore x = 2 \text{ and } y = 3$

b

$3x + 5y = 17$ (1)
$3x - 4y = -19$ (2)

$(1) - (2)$

$5y - (-4y) = 17 - (-19)$
$5y + 4y = 17 + 19$
$9y = 36$
$y = 4$

Sub $y = 4$ into (1)

$3x + 5y = 17$
$3x + 5(4) = 17$
$3x + 20 = 17$
$3x = -3$
$x = -1$

$\therefore x = -1 \text{ and } y = 4$

c

$6a - 4b = 14$ (1)
$-2a - 4b = -10$ (2)

$(1) - (2)$

$6a - (-2a) = 14 - (-10)$
$8a = 24$
$a = 3$

Sub $a = 3$ into (1)

$6a - 4b = 14$
$6(3) - 4b = 14$
$18 - 4b = 14$
$-4b = -4$
$b = 1$

$\therefore a = 3 \text{ and } b = 1$

d

$7a + 6b = -8$ (1)
$4a - 6b = -14$ (2)

$(1) + (2)$

$7a + 4a = -8 + (-14)$
$11a = -22$
$a = -2$

Sub $a = -2$ into (2)

$4a - 6b = -14$
$4(-2) - 6b = -14$
$-8 - 6b = -14$
$-6b = -6$
$b = 1$

$\therefore a = -2 \text{ and } b = 1$

1 e

$11p - 4q = 2 \quad (1)$
$7p - 4q = -6 \quad (2)$

$(1) - (2)$

$11p - 7p = 2 - (-6)$
$4p = 8$
$p = 2$

Sub $p = 2$ into (1)

$11p - 4q = 2$
$11(2) - 4q = 2$
$22 - 4q = 2$
$20 = 4q$
$q = 5$

$\therefore p = 2 \ and \ q = 5$

f

$3p - 5q = 6 \quad (1)$
$-2p - 5q = 46 \quad (2)$

$(2) - (1)$

$-2p - 3p = 46 - 6$
$-5p = 40$
$p = -8$

Sub $p = -8$ into (1)

$3p - 5q = 6$
$3(-8) - 5q = 6$
$-24 - 5q = 6$
$-5q = 30$
$q = -6$

$\therefore p = -8 \ and \ q = -6$

g

$7x - 3y = -16 \quad (1)$
$-7x + 2y = 6 \quad (2)$

$(1) + (2)$

$-3y + 2y = -16 + 6$
$-y = 10$
$y = 10$

Sub $y = 10$ into (2)

$-7x + 2y = 6$
$-7x + 2(10) = 6$
$-7x + 20 = 6$
$-7x = -14$
$x = 2$

$\therefore x = 2 \ and \ y = 10$

h

$-3x + 4y = 14 \quad (1)$
$5x + 4y = -34 \quad (2)$

$(1) - (2)$

$-3x - 5x = 14 - (-34)$
$-8x = 18 + 34$
$-8x = 48$
$x = -6$

Sub $x = -6$ into (1)

$-3x + 4y = 14$
$-3(-6) + 4y = 14$
$18 + 4y = 14$
$4y = -4$
$y = -1$

$\therefore x = -6 \ and \ y = -1$

2 **a**

$2x + 3y = 9$ (1)
$4x + 5y = 15$ (2)

$2 \times (1)$

$2 \times (2x + 3y = 9)$
$4x + 6y = 18$ (3)

$4x + 5y = 15$ (2)
$4x + 6y = 18$ (3)

$(3) - (2)$

$6y - 5y = 18 - 15$
$y = 3$

Sub $y = 3$ into (1)

$2x + 3y = 9$
$2x + 3(3) = 9$
$2x + 9 = 9$
$2x = 0$
$x = 0$

$\therefore x = 0 \text{ and } y = 3$

b

$3x + 7y = 1$ (1)
$9x - 4y = -22$ (2)

$3 \times (1)$

$3 \times (3x + 7y = 1)$
$9x + 21y = 3$ (3)

$9x - 4y = -22$ (2)
$9x + 21y = 3$ (3)

$(3) - (2)$

$21y - (-4y) = 3 - (-22)$
$21y + 4y = 3 + 22$
$25y = 25$
$y = 1$

Sub $y = 1$ into (1)

$3x + 7y = 1$
$3x + 7(1) = 1$
$3x + 7 = 1$
$3x = -6$
$x = -2$

$\therefore x = -2 \text{ and } y = 1$

2 **c**

$12a + 3b = 3$ (1)
$-4a - 5b = 11$ (2)

$3 \times (2)$

$3 \times (-4a - 5b = 11)$
$-12a - 15b = 33$ (3)

$12a + 3b = 3$ (1)
$-12a - 15b = 33$ (3)

$(3) + (1)$

$-15b + 3b = 33 + 3$
$-12b = 36$
$b = -3$

Sub $b = -3$ into (1)

$12a + 3b = 3$
$12a + 3(-3) = 3$
$12a - 9 = 3$
$12a = 12$
$a = 1$

$\therefore a = 1 \text{ and } b = -3$

d

$7a - 6b = -9$ (1)
$2a + 12b = -30$ (2)

$2 \times (1)$

$2 \times (7a - 6b = -9)$
$14a - 12b = -18$ (3)

$2a + 12b = -30$ (2)
$14a - 12b = -18$ (3)

$(3) + (2)$

$14a + 2a = -18 + (-30)$
$16a = -18 - 30$
$16a = -48$
$a = -3$

Sub $a = -3$ into (1)

$7a - 6b = -9$
$7(-3) - 6b = -9$
$-21 - 6b = -9$
$-6b = 12$
$b = -2$

$\therefore a = -3 \text{ and } b = -2$

2 **e**

$5p + 2q = -7$ (1)
$2p - 20q = 18$ (2)

$10 \times (1)$

$10 \times (5p + 2q = -7)$
$50p + 20q = -70$ (3)

$2p - 20q = 18$ (2)
$50p + 20q = -70$ (3)

$(2) + (3)$

$2p + 50p = 18 + (-70)$
$52p = -52$
$p = -1$

Sub $p = -1$ into (1)

$5p + 2q = -7$
$5(-1) + 2q = -7$
$-5 + 2q = -7$
$2q = -2$
$q = -1$

$\therefore p = -1 \text{ and } q = -1$

f

$2p - 3q = 0$ (1)
$5p + 15q = 45$ (2)

$5 \times (1)$

$5 \times (2p - 3q = 0)$
$10p - 15q = 0$ (3)

$5p + 15q = 45$ (2)
$10p - 15q = 0$ (3)

$(3) + (2)$

$10p + 5p = 45$
$15p = 45$
$p = 3$

Sub $p = 3$ into (1)

$2p - 3q = 0$
$2(3) - 3q = 0$
$6 - 3q = 0$
$6 = 3q$
$q = 2$

$\therefore p = 3 \text{ and } q = 2$

2 **g** $-2x + 5y = -10$ (1)
$8x - 3y = 6$ (2)

$4 \times (1)$

$4 \times (-2x + 5y = -10)$
$-8x + 20y = -40$ (3)

$8x - 3y = 6$ (2)
$-8x + 20y = -40$ (3)

$(3) + (2)$

$20y + (-3y) = -40 + 6$
$20y - 3y = -34$
$20y - 3y = -34$
$17y = -34$
$y = -2$

Sub $y = -2$ into (2)

$8x - 3y = 6$
$8x - 3(-2) = 6$
$8x + 6 = 6$
$8x = 0$
$x = 0$

$\therefore x = 0 \ and \ y = -2$

h $-5x - 6y = 4$ (1)
$3x - 2y = 20$ (2)

$3 \times (2)$

$3 \times (3x - 2y = 20)$
$9x - 6y = 60$ (3)

$-5x - 6y = 4$ (1)
$9x - 6y = 60$ (3)

$(3) - (1)$

$9x - (-5x) = 60 - 4$
$9x + 5x = 56$
$14x = 56$
$x = 4$

Sub $x = 4$ into (2)

$3x - 2y = 20$
$3(4) - 2y = 20$
$12 - 2y = 20$
$-2y = 8$
$y = -4$

$\therefore x = 4 \ and \ y = -4$

3 **a**

$3x + 2y = 7 \qquad (1)$
$2x + 5y = 12 \qquad (2)$

$2 \times (1)$
$3 \times (2)$

$2 \times (3x + 2y = 7)$
$3 \times (2x + 5y = 12\ )$

$6x + 4y = 14 \qquad (3)$
$6x + 15y = 36 \qquad (4)$

$(4) - (3)$

$15y - 4y = 36 - 14$
$11y = 22$
$y = 2$

Sub $y = 2$ into (2)

$2x + 5y = 12$
$2x + 5(2) = 12$
$2x + 10 = 12$
$2x = 2$
$x = 1$

$\therefore x = 1 \textit{ and } y = 2$

b

$-3x + 2y = -7 \qquad (1)$
$4x - 5y = 7 \qquad (2)$

$5 \times (1)$
$2 \times (2)$

$5 \times (-3x + 2y = -7)$
$2 \times (4x - 5y = 7)$

$-15x + 10y = -35 \qquad (3)$
$8x - 10y = 14 \qquad (4)$

$(4) + (3)$

$8x + (-15x) = 14 + (-35)$
$8x - 15x = 14 - 35$
$-7x = -21$
$x = 3$

Sub $x = 3$ into (2)

$4x - 5y = 7$
$4(3) - 5y = 7$
$12 - 5y = 7$
$12 - 7 = 5y$
$5y = 5$
$y = 1$

$\therefore x = 3 \textit{ and } y = 1$

3 **c**

$4a + 3b = -5$ (1)
$3a - 2b = 9$ (2)

$2 \times (1)$
$3 \times (2)$

$2 \times (4a + 3b = -5)$
$3 \times (3a - 2b = 9)$

$8a + 6b = -10$ (3)
$9a - 6b = 27$ (4)

$(4) + (3)$

$9a + 8a = 27 + (-10)$
$17a = 17$
$a = 1$

Sub $a = 1$ into (2)

$3a - 2b = 9$
$3(1) - 2b = 9$
$3 - 2b = 9$
$-2b = 6$
$b = -3$

$\therefore a = 1 \; and \; b = -3$

d

$-3a + 4b = -2$ (1)
$-4a - 5b = 18$ (2)

$4 \times (1)$
$3 \times (2)$

$4 \times (-3a + 4b = -2)$
$3 \times (-4a - 5b = 18)$

$-12a + 16b = -8$ (3)
$-12a - 15b = 54$ (4)

$(3) - (4)$

$16b - (-15b) = -8 - 54$
$16b + 15b = -62$
$31b = -62$
$b = -2$

Sub $b = -2$ into (1)

$-3a + 4b = -2$
$-3a + 4(-2) = -2$
$-3a - 8 = -2$
$-3a = 6$
$a = -2$

$\therefore a = -2 \; and \; b = -2$

3 **e**

$7p - 3q = 5$ (1)
$-5p + 4q = -11$ (2)

$4 \times (1)$
$3 \times (2)$

$4 \times (7p - 3q = 5)$
$3 \times (-5p + 4q = -11)$

$28p - 12q = 20$ (3)
$-15p + 12q = -33$ (4)

$(4) + (3)$

$-15p + 28p = -33 + 20$
$13p = -13$
$p = -1$

Sub $p = -1$ into (1)

$7p - 3q = 5$
$7(-1) - 3q = 5$
$-7 - 3q = 5$
$-3q = 12$
$q = -4$

$\therefore p = -1 \text{ and } q = -4$

f

$-2p + 5q = 2$ (1)
$9p - 3q = 30$ (2)

$9 \times (1)$
$2 \times (2)$

$9 \times (-2p + 5q = 2)$
$2 \times (9p - 3q = 30)$

$-18p + 45q = 18$ (3)
$18p - 6q = 60$ (4)

$(3) + (4)$

$45q + (-6q) = 18 + 60$
$45q - 6q = 78$
$39q = 78$
$q = 2$

Sub $q = 2$ into (1)

$9p - 3q = 30$
$9p - 3(2) = 30$
$9p - 6 = 30$
$9p = 36$
$p = 4$

$\therefore p = 4 \text{ and } q = 2$

3 **g**

$-8x - 3y = 7$ (1)
$5x + 2y = -5$ (2)

$2 \times (1)$
$3 \times (2)$

$2 \times (-8x - 3y = 7)$
$3 \times (5x + 2y = -5)$

$-16x - 6y = 14$ (3)
$15x + 6y = -15$ (4)

$(3) + (4)$

$-16x + 15x = 14 + (-15)$
$-x = 14 - 15$
$-x = -1$
$x = 1$

Sub $x = 1$ into (2)

$5x + 2y = -5$
$5(1) + 2y = -5$
$5 + 2y = -5$
$2y = -10$
$y = -5$

$\therefore x = 1 \ and \ y = -5$

h

$-3x - 7y = -9$ (1)
$-5x - 4y = 8$ (2)

$5 \times (1)$
$3 \times (2)$

$5 \times (-3x - 7y = -9)$
$3 \times (-5x - 4y = 8)$

$-15x - 35y = -45$ (3)
$-15x - 12y = 24$ (4)

$(4) - (3)$

$-12y + 35y = 24 + 45$
$23y = 69$
$y = 3$

Sub $y = 3$ into (2)

$-5x - 4y = 8$
$-5x - 4(3) = 8$
$-5x - 12 = 8$
$-5x = 20$
$x = -4$

$\therefore x = -4 \ and \ y = 3$

Challenge Yourself

1 **a**

$3 + 4x - 3y = 0$
$4 - 2y + 3x = 0$

$4x - 3y = -3$ (1)
$3x - 2y = -4$ (2)

$2 \times (1)$
$3 \times (2)$

$2 \times (4x - 3y = -3)$
$3 \times (3x - 2y = -4)$

$8x - 6y = -6$ (3)
$9x - 6y = -12$ (4)

$(4) - (3)$

$9x - 8x = -12 - (-6)$
$x = -12 + 6$
$x = -6$

Sub $x = -6$ into (2)

$3x - 2y = -4$
$3(-6) - 2y = -4$
$-18 - 2y = -4$
$-2y = 14$
$y = -7$

$\therefore x = -6 \text{ and } y = -7$

b

$2x = 15 + 5y$
$-7y = 6x - 23$

$2x - 5y = 15$ (1)
$-6x - 7y = -23$ (2)

$3 \times (1)$

$3 \times (2x - 5y = 15)$ (3)
$6x - 15y = 45$

$-6x - 7y = -23$ (2)
$6x - 15y = 45$ (3)

$(2) + (3)$

$-7y - 15y = -23 + 45$
$-22y = 22$
$y = -1$

Sub $y = -1$ into (1)

$2x - 5y = 15$
$2x - 5(-1) = 15$
$2x + 5 = 15$
$2x = 10$
$x = 5$

$\therefore x = 5 \text{ and } y = -1$

1 c

$$\frac{2m-5n}{2} = 16 \quad (1)$$
$$2m + 3n = 0 \quad (2)$$

Get rid of fraction (1)

$$\frac{2m-5n}{2} = 16$$
$$2m - 5n = 32 \quad (3)$$

$$2m + 3n = 0 \quad (2)$$
$$2m - 5n = 32 \quad (3)$$

(2) − (3)

$$3n - (-5n) = 0 - 32$$
$$3n + 5n = -32$$
$$8n = -32$$
$$n = -4$$

Sub $n = -4$ into (2)

$$2m + 3n = 0$$
$$2m + 3(-4) = 0$$
$$2m - 12 = 0$$
$$2m = 12$$
$$m = 6$$

$\therefore m = 6 \text{ and } n = -4$

d

$$\frac{2p}{3} + 3q = -4 \quad (1)$$
$$5p - 9q = 33 \quad (2)$$

$3 \times (1)$

$$3 \times \left(\frac{2p}{3} + 3q = -4\right)$$
$$2p + 9q = -12 \quad (3)$$

$$5p - 9q = 33 \quad (2)$$
$$2p + 9q = -12 \quad (3)$$

(2) + (3)

$$5p + 2p = 33 + (-12)$$
$$7p = 33 - 12$$
$$7p = 21$$
$$p = 3$$

Sub $p = 3$ into (2)

$$5p - 9q = 33$$
$$5(3) - 9q = 33$$
$$15 - 9q = 33$$
$$-9q = 18$$
$$q = -2$$

$\therefore p = 3 \text{ and } q = -2$

1 e

$$\frac{x+2}{2}+\frac{y-1}{3}=1 \quad (1)$$

$$5x+2y=-6 \quad (2)$$

Get rid of fractions (1)

$$\frac{x+2}{2}+\frac{y-1}{3}=1$$

$$\frac{3(x+2)+2(y-1)}{6}=1$$

$$\frac{3x+6+2y-2}{6}=1$$

$$\frac{3x+2y+4}{6}=1$$

$$3x+2y+4=6$$

$$3x+2y=2 \quad (3)$$

$$5x+2y=-6 \quad (2)$$
$$3x+2y=2 \quad (3)$$

(2) − (3)

$$5x-3x=-6-3$$
$$2x=-8$$
$$x=-4$$

Sub $x=-4$ into (2)

$$5x+2y=-6$$
$$5(-4)+2y=-6$$
$$-20+2y=-6$$
$$2y=14$$
$$y=7$$

$$\therefore x=-4 \text{ and } y=7$$

1 f $\frac{2x}{3} + \frac{y}{2} = \frac{1}{6}$ (1)

$\frac{x}{5} + \frac{3y}{2} = -\frac{2}{5}$ (2)

Get rid of fractions (1)

$\frac{2x}{3} + \frac{y}{2} = \frac{1}{6}$

$\frac{4x + 3y}{6} = \frac{1}{6}$

$4x + 3y = 1$ (3)

Get rid of fractions (2)

$\frac{x}{5} + \frac{3y}{2} = -\frac{2}{5}$

$\frac{2x + 15y}{10} = -\frac{2}{5}$

$2x + 15y = -4$ (4)

$4x + 3y = 1$ (3)
$2x + 15y = -4$ (4)

$2 \times (4)$

$2 \times (2x + 15y = -4)$
$4x + 30y = -8$ (5)

$4x + 3y = 1$ (3)
$4x + 30y = -8$ (5)

$(5) - (3)$

$30y - 3y = -8 - 1$
$27y = -9$

$y = -\frac{9}{27} = -\frac{1}{3}$

Sub $y = -\frac{1}{3}$ into (3)

$4x + 3y = 1$
$4x + 3\left(-\frac{1}{3}\right) = 1$

$4x - 1 = 1$
$4x = 2$

$x = \frac{2}{4} = \frac{1}{2}$

$\therefore x = \frac{1}{2}$ *and* $y = -\frac{1}{3}$

Was the answer book any good?

I genuinely hope this answer book helped you learn.

And I'd love to know if it did.

So, if you don't mind, can you write a review?

Tell me what you think.

Did it help?

Were the steps clear?

Any mistakes?

Better not be.

I triple checked everything.

But seriously, let me know.

Anything.

Good or bad.

Appreciate it.

Made in United States
North Haven, CT
06 November 2026

21884525R00098